Polymer Materials for Electronic Applications

Eugene D. Feit, EDITOR

Harris Semiconductor Group

Cletus W. Wilkins, Jr., EDITOR

Bell Laboratories

Based on a symposium

sponsored by the Division of

Organic Coatings and Plastics Chemistry

at the Second Chemical Congress

of the North American Continent

(180th ACS National Meeting),

Las Vegas, Nevada,

August 26–27, 1980.

ACS SYMPOSIUM SERIES **184**

AMERICAN CHEMICAL SOCIETY

WASHINGTON, D. C.   1982

Library of Congress Cataloging in Publication Data

Polymer materials for electronic applications.
(ACS symposium series, ISSN 0097–6156; 184)

Includes bibliographies and index.

1. Electronics—Materials—Congresses. 2. Polymers
and polymerization—Congresses.
I. Feit, Eugene D., 1935– . II. Wilkins, Cletus W.,
1945– . III. American Chemical Society. Division of
Organic Coatings and Plastics Chemistry. IV. Series.

TK7871.15.P6P64   1982   621.381′028   82–1670
ISBN 0–8412–0715–1

FOREWORD

The ACS Symposium Series was founded in 1974 to provide a medium for publishing symposia quickly in book form. The format of the Series parallels that of the continuing Advances in Chemistry Series except that in order to save time the papers are not typeset but are reproduced as they are submitted by the authors in camera-ready form. Papers are reviewed under the supervision of the Editors with the assistance of the Series Advisory Board and are selected to maintain the integrity of the symposia; however, verbatim reproductions of previously published papers are not accepted. Both reviews and reports of research are acceptable since symposia may embrace both types of presentation.

CONTENTS

PREFACE

The 180th National Meeting of the American Chemical Society in August, 1980 witnessed the fourth symposium on Polymeric Materials for Electronic Applications. This symposium marks the first time that full papers are presented in the ACS SYMPOSIUM SERIES. The first symposium was held in Dallas in April, 1973 at the 165th ACS Meeting as part of L. H. Princen's Second Symposium on Scanning Electron Microscopy on Polymers and Coatings. By August of 1975 at the 170th ACS Meeting in Chicago, the symposium stood by itself with 17 papers. The third symposium occurred in 1977, again in Chicago, at the 174th ACS Meeting. The proceedings of all four symposia were recorded in preprint volumes of the Division of Organic Coatings and Plastics Chemistry.

The size and scope of the symposium have changed over the years. In 1973 only seven papers were presented, all of which reported on electron-beam resists. The number of papers rose to 17, then to 20, and then to 24, but the number describing electron-beam resists has dropped steadily; in this symposium, electron-beam resists constituted only a minority of the papers. This broadening of the topical base for the symposium indicates the level to which chemistry penetrates the electronics industry.

Organic coatings are used in the electronics industry, both as resists and as encapsulants and insulating, intermediate dielectrics, and perhaps in the not too distant future, as conducting elements. The goal of the symposium organizers has been to emphasize the chemical aspects of these materials and their uses. This emphasis concerns mechanisms of formation and utilization, chemical stability, change, and reliability, as consequences of chemical composition and reactions.

The organization of this volume closely follows the organization of the symposium. Approximately half of the chapters involve organic or organometallic materials for image transfer. The growth of very large scale integration in microelectronics would not be possible without advances in sensitivity and resolution of electron- and photon-resists, and this is a common theme throughout the book. Approximately a quarter of the chapters deal with the use of polyimides as insulating dielectrics. (A measure of the importance of this area is the fact that it was among the best attended sessions of the entire Organic Coatings and Plastics Division.) The remaining chapters concern a diversity of applications including encapsulants, synthesis and properties of branched epoxy resins, and thermal degradation of polymers for molded integrated circuit devices.

As symposium chairmen, we sincerely appreciate the support of all the authors and participants for this well-received symposium. We also hope that our readers will find this volume instructive and timely.

E. D. FEIT
Harris Semiconductor Group
Melbourne, Florida 32901

C. W. WILKINS, JR.
Bell Laboratories
Murray Hill, New Jersey 07974

December 1981.

A Sensitive Positive-Working Cross-Linked Methacrylate Electron Resist

E. D. ROBERTS

Philips Research Laboratories, Redhill, Surrey, England

In earlier publications (1,2), a cross-linked methacrylate positive-working electron resist was described. The resist consists of a mixture of two copolymers- poly-(methyl methacrylate -co-methacrylic acid) and poly-(methyl methacrylate-co-methacryloyl chloride), both of which, having straight chain structures, are soluble in organic solvents. A solution of the mixture is used to apply a film to a substrate by the normal spinning process. When the film is subsequently baked, the carboxyl groups react with the chlorocarbonyl groups to form carboxylic acid anhydride cross-links.

The cross-linked polymer is, like all cross-linked materials, completely insoluble in all organic solvents. The chemical bonds joining the anhydride group to the main methacrylate chain are particularly susceptible to rupture by ionising radiations, so upon exposure to an electron beam they are broken and the straight chain structure is restored. The irradiated polymer thus becomes soluble again and can be selectively dissolved to give a positive image in the resist film. Methyl isobutyl ketone is normally used as the developing solvent.

The system is capable of almost infinite variation, for the proportion of cross-links in the structure can be varied over a wide range. Firstly, it can be changed by altering the proportion of potential cross-linking groups (carboxyl and

0097-6156/82/0184-0001$05.00/0

chlorocarbonyl) in the copolymers and choosing conditions to
ensure that most or all of these react. Alternatively, in a
system of particular composition, the proportion of potential
cross-linking groups which actually react can be changed by
altering the conditions under which the cross-links are formed -
that is, by changing the curing conditions.

The curves in Figure 1 show some of the variety of properties
which can be obtained from these systems. It was shown
earlier (1,2) that these resists are more stable to thermal
deformation than is poly-(methyl methacrylate)(PMMA), at least
for films cured at 175°C. At this curing temperature, presumably,
the maximum possible degree of cross-linking has been achieved
and the sensitivity is a minimum. The relatively low
sensitivity is not necessarily disadvantageous in resists intended
for use in equipment such as electron image projector systems
(3,4). For use with pattern generators, (5,6), however, it is
desirable to have a resist of higher sensitivity (7). The
earlier publications (1,2) described mainly the examination
and use of relatively highly cross-linked films which had been
cured at 175°C. This paper describes some of the properties
obtained when the resists are very lightly cross-linked, by
curing at temperatures in the range 100-125°C. Under these
conditions, much greater sensitivity is obtained, though more
careful control of operating conditions is needed to obtain
reproducible results.

Properties of sensitive cross-linking resists

The copolymer system used in the work described here is our
standard cross-linking resist, comprising the mixture of the two
copolymers specified above, each containing 10 mol% of the
comonomer which provides the potential cross-linking groups. The
degree of cross-linking introduced has been restricted by curing
at temperatures within the range 100-125°C, and Table I shows the
sensitivities of films cured under these conditions. At these
relatively low curing temperatures, the sensitivity changes fairly
rapidly with curing temperature, so it is necessary to maintain
that temperature within about half a degree of the intended value
if reproducible results are to be obtained. However, this does
not present an insuperable difficulty.

It is apparent from Table I that the sensitivity of lightly
cross-linked resist films is more dependent upon film thickness
than is that of fully-cured films (i.e. films cured at 150-175°C).
It seems that the solvent may have some influence upon
sensitivity. The values in Table I were obtained on films spun
from solution in ethoxyethyl acetate and even fully cross-linked
films show some variation of sensitivity with film thickness.
This is in contrast to our earlier work in which films spun from
solution in methyl isobutyl ketone had sensitivity 40μC/cm^2 at
any film thickness from about 0.3 to 2μm. It must be more

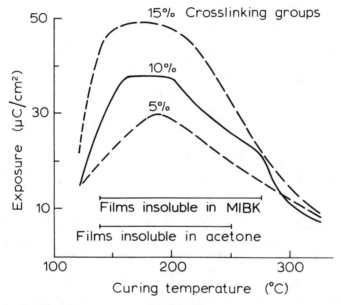

Figure 1. *Variation of sensitivity of cross-linked methacrylate resists with curing temperature (8).*

Table I

Variation of Properties of Cross-Linked Methacrylate Resists
with Curing Conditions

| Curing Temp. °C | Curing Time Mins | Dev.Time at 22-23°C Mins | Sensitivity µC/cm^2 | | | | Typical loss of thickness during development at 22-23°C. µm |
| | | | ~0.5µm thickness | | ~1.0µm thickness | | |
			10kV	20kV	10kV	20kV	
150-200 Optimum 175	15	2-4	35-40	110-115	50	–	None
123	45	4	8	–	–	–	None
120	45	4	6	26	–	–	0.03
115	45	4	6	–	–	–	0.04
110	60	5	3	16	11	–	0.05
105	60	5	3	12	7	26	0.12
100	60	5	2	4	6	–	0.29

difficult to eliminate ethoxyethyl acetate from the film but it
is not clear why this should affect the sensitivity in the way
it appears to.

The increase in sensitivity is not obtained without some
sacrifice in other properties. During the development process,
there is some loss of thickness in unexposed areas of the film,
and this loss becomes greater as the sensitivity is increased.
The actual loss in thickness is almost independent of the original
film thickness and the curves in Figure 2 show the rate at which
this thickness changes, together with the rate at which PMMA
dissolves. Of course, the cross-linked films never dissolve
completely, and during a normal development process involving
five minutes immersion, the loss is generally too small to cause
any difficulty, at least with films cured at 105°C or at higher
temperatures. Many users appear to achieve satisfactory results
even with films cured at 100°C. The last column of Table I
shows typical values for the thickness loss during development of
films cured at different temperatures.

Both the thickness loss and sensitivity depend upon
developing time and temperature, and the conditions specified
in Table I were chosen to give generally a reasonable compromise
between two opposing requirements. Table II shows the kind of
variations which can occur in lightly cross-linked resists
when the developer temperature is changed, so it is clear that
this temperature also needs to be controlled to within about 1°C
to achieve good reproducibility when using them.

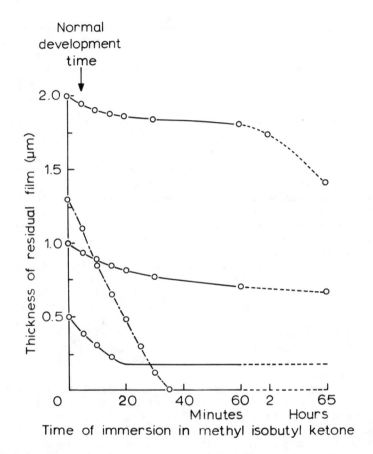

Figure 2.   Rate of dissolution of methacrylate films. Key: · – · –, PMMA; — — —, cross-linked resist cured 1 h at 105°C.

Table II

Variation of Cross-linked Methacrylate Resist
Properties with Developing Conditions

Developer Temperature °C	Sensitivity at 10kV μC/cm^2		Loss of thickness during development μm	
	∿0.5μm thickness	∿1.0μm thickness	∿0.5μm thickness	∿1.0μm thickness
20	5	8	0.11	0.08
25	2	7	0.13	0.13

All resists were cured for 1 hour at 105°C and developed in
methyl isobutyl ketone for 5 minutes.

There is one other source of potential difficulty, though
this has only arisen when seeking to obtain the highest
sensitivity by curing at 100°C or less, and even then only
when using films at least 1μm thick. This is the appearance
during the pattern development stage of cracks in the film.
During a normal development process, the cracks appear only in
the immediate vicinity of the exposed area and almost invariably
originate from the corners of patterns as shown in Figure 3.
If the film is left to soak in the developer for some time,
during which more of it dissolves (Figure 2), the cracks
disappear again. This suggests that they are only surface cracks.
No cracking has been observed in ½μm thick films cured at 100°C
nor in films of any thickness cured at 105°C or above. These
observations are consistent with the cracks being produced by
surface swelling during development followed by shrinkage during
the drying stage.

The ultimate cause of the effect is believed to be partial
breakdown of the cross-linked structure by scattered electrons
acting upon the surface of the film, making its surface layers
more susceptible to swelling during development. Thicker films
need higher exposures than thin ones do, so the exposure and
consequent breakdown by scattered electrons is proportionately
higher, and presumably is sufficient to make the film prone to
swelling during development in this case. If the exposure is
increased further, the cracking becomes more extensive though
at still higher exposures it is not apparent at all. In this
case, it appears that the surface cross-links have been broken
sufficiently to render the surface layer truly soluble. Indeed,
the residual film is thinner in the immediate vicinity of the
exposed area.

This explanation is supported by the fact that if a more
powerful solvent such as acetone is used as the developer,

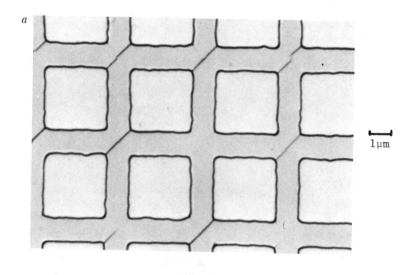

Figure 3. *Cracks in thick cross-linked methacrylate resist films cured at 100°C for 1 h, exposed at 7 μC/cm² at 10 kV and developed for 5 min in methyl isobutyl ketone. Key: a, optical micrograph and b, scanning electron micrograph (viewed at 60°).*

exactly analogous behaviour is observed, though corresponding
effects occur at rather higher curing temperatures. This
reflects the ability of the stronger solvent to produce
swelling in more highly cross-linked structures than the weaker
solvent, methyl isobutyl ketone, can.

Thermal stability of lightly cross-linked films

The patterns which are shown in Figures 3 to 7, were made
by exposing the films to a flood beam of electrons of energy
10kV, using a fine copper gauze as a contact mask. After
development, the film patterns were heated to various
temperatures, cleaved and examined for signs of deformation.
Figure 4 shows that the characteristic undercut profile, which
in similarly treated PMMA disappears on heating between 90 and
$100^{\circ}C$ (2), is still present in the lightly cross-linked film
which has been heated for 15 minutes at $130^{\circ}C$. At higher
temperatures the undercut profile does disappear but there does
not seem to be any change in lateral dimensions and edge
details are not destroyed to the extent that they are in PMMA.
Figure 5a is a pattern in a film cured for 1 hour at $100^{\circ}C$
and exposed at $2\mu C/cm^2$, the residual film thickness being about
a third of a micron. Figure 5b is the same film after being
heated for 20 minutes at $175^{\circ}C$ and the edge detail appears to
be unchanged after this treatment. Figures 6a and 6b are
patterns in a PMMA film which is also one third of a micron
thick after development and has been heated in the same way
(Figure 6b). The flow of the polymer film is much more
extensive and has destroyed all the detail in the pattern edges.
If the user is prepared to sacrifice some sensitivity,
much greater thermal stability in developed patterns can be
attained. Figure 7 shows a film which was cured for 45 minutes
at $125^{\circ}C$, and exposed at $12\mu C/cm^2$ at 10kV. Under these
conditions, unexposed parts are completely insoluble during the
development process (Table I). The pictures show the edge
profiles of a pattern which, after development, has been heated
at various elevated temperatures. During this heat treatment,
no flow of the undercut edge has occurred.

Discussion on thermal stability. The most unexpected
feature of this investigation is the good thermal stability of
these lightly cross-linked resists, and it is particularly
striking in the film pattern which has been heated at $250^{\circ}C$
(Figure 7c) apparently without changing. The same area of a
number of samples has been examined after successively
increased heat treatment. The only change which can be seen
is the appearance of undulations in the surface, like that
which can be seen in Figure 7b. However, these occur only
within areas which have been scanned during the previous
electron microscope examination and so certainly result from

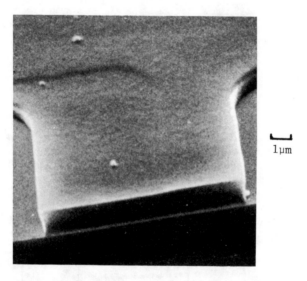

1 μm

Figure 4.   Edge profile of a pattern in cross-linked methacrylate resist film cured at 105°C and exposed at 8 μC/cm² at 10 kV after heating for 15 min at 130°C (8).

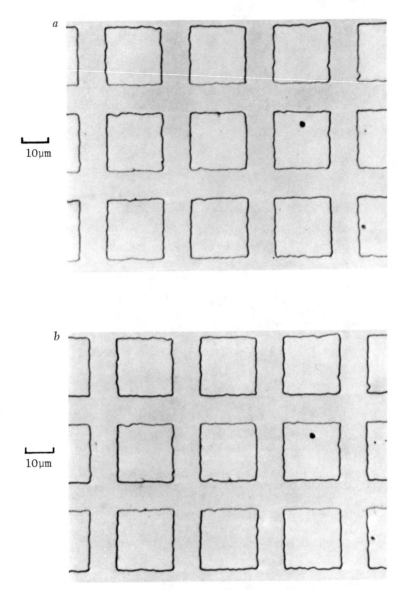

Figure 5.   Optical micrographs of patterns in cross-linked methacrylate resist film cured at 100°C for 1 h, exposed at 2 μC/cm² at 10 kV, and developed for 5 min in methyl isobutyl ketone (8). Key: a, as developed and b, after heating for 20 min at 175°C.

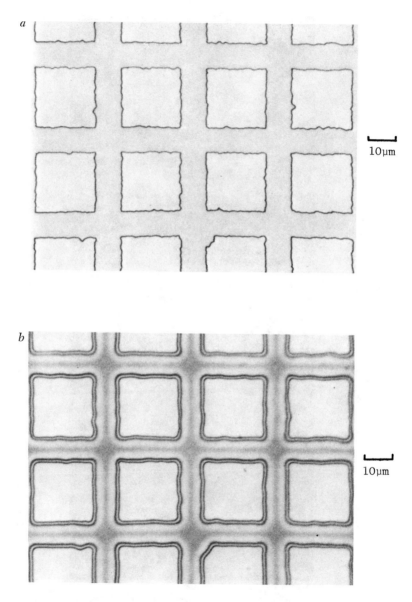

Figure 6.   Optical micrographs of patterns in PMMA exposed at 50 μC/cm² at 10 kV (8). Key: a, as developed and b, after heating for 20 min at 175°C.

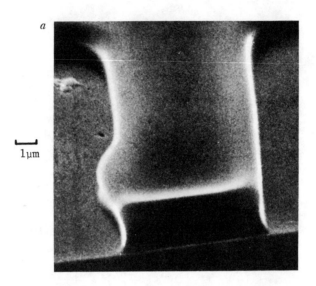

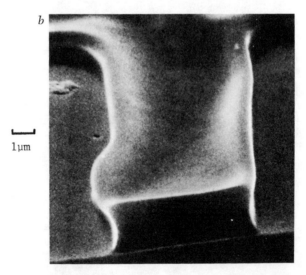

Figure 7. Edge profiles of patterns in cross-linked methacrylate resist film cured at 125° C for 45 min and exposed at 12 μC/cm² at 10 kV (viewed at 60°). Key: a, after heating for 20 min at 175°C and b, after further heating for 20 min at 200°C. Continued.

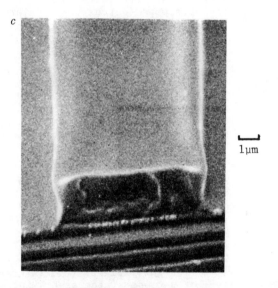

Figure 7.   Edge profiles of patterns in cross-linked methacrylate resist film cured at 125° C for 45 min and exposed at 12 μC/cm² at 10 kV (viewed at 60°). Key: c, similar pattern as (a) and (b) after heating for 20 min at 250°C (8).

exposure during that examination. In all the samples examined, no changes in profile of the patterns were seen to have occurred as a result of the heat treatment.

It was shown earlier (2) that when the same resist is fully cured, by baking at 175°C, film patterns show signs of flow as a change in slope of the edge profile after heating at 175°C. This was, and still is, believed to be due to breakdown of the cross-linked structure just in the edges of the nominally unirradiated areas, caused by electrons scattered from adjacent irradiated areas. In principle, this can still happen in the sensitive form of this resist although the exposure levels are lower. However, because the film is now only lightly cross-linked before exposure, a number of the potential cross-linking groups (carboxyl and chlorocarbonyl) will still be present in their original forms throughout the film remaining on the substrate after exposure and development. Even though the original cross-linked structure may have been broken to some extent by scattered electrons, the potential cross-linking groups still in the film can react to form new cross-links during any subsequent heating process. Presumably, this occurs sufficiently rapidly and is extensive enough to enable the film to resist deformation at higher temperatures.

In films cured initially at 175°C, determination by infra-red spectroscopy of the concentration of anhydride groups indicated that all potential cross-linking groups had reacted. In this case, therefore, no groups remain to form new cross-links upon heating after irradiation, and some flow can occur where the structure has been partially degraded by scattered electrons.

Supporting evidence for this hypothesis of post-exposure cross-linking came in other experiments in which resist films cured initially at 100°C and subjected to various exposures, were then post-baked at slightly higher temperatures before development. The purpose of the experiment was actually to discover if the rather high loss of thickness which occurred during development of patterns in films cured at 100°C could be reduced, and indeed it was. Unfortunately, the sensitivity was also reduced, and the overall effect was as if the initial curing had been performed actually at the post-baking temperature. In this case, further cross-linking was clearly taking place during the post-baking process, even in the irradiated areas, for these did not develop properly unless the exposure exceeded the sensitivity corresponding to the post-bake temperature.

Uses of sensitive cross-linked resists

The resists can be used for defining fine details, and Figure 8a and 8b show some produced by a pattern generator in a film which was cured at 100°C and exposed at $3.2\mu C/cm^2$ at 20kV. The lines are 1, 1.5 and 3μm wide.

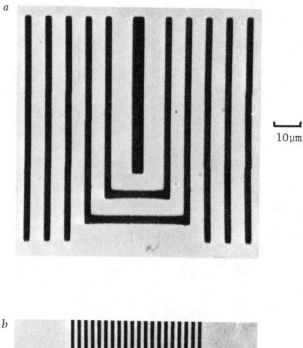

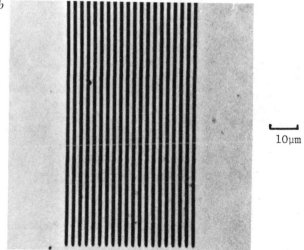

Figure 8. Fine lines in cross-linked methacrylate resist film cured at 100°C for 1 h, exposed at 3.2 μC/cm² at 20 kV, and developed for 5 min in methyl isobutyl ketone. Key: a, 1.5-μm and 3-μm wide lines (8) and b, 1-μm wide lines.

The resist has been used as a mask in wet etching and in lift-off processes, and more recently in etching chromium films in a chlorine-oxygen-helium plasma. In the latter, the etch rates have ranged from 4 to 5.5nm/min at 100W power in a barrel type reactor. Chromium etches at about 6.5nm/min under these conditions. The etch rate of the resist appears to be independent of the degree to which it has been cured before exposure, so the sensitive form described here is just as effective a mask as the highly cross-linked resists described earlier, at least in the chromium etching process.

Conclusions

Sensitive cross-linking positive-working resists can be made by restricting the degree of cross-linking introduced into the resist films. This can be done by limiting the proportion of potential cross-linking groups in the constituent copolymers, though considerably greater thermal stability is obtained if it is done rather by restricting the curing temperature of a system containing a relatively high proportion of potential cross-linking groups. By selecting suitable curing conditions, one resist mixture can be used to obtain sensitivities appropriate to the type of electron beam equipment in which it will be used - image projectors or pattern generators.

Acknowledgments

The author thanks the Directors of Philips Research Laboratories for permission to publish this paper. He is also grateful to many of his colleagues in the Laboratories for stimulating and helpful discussions, particularly concerning the use of the materials described.

Literature Cited

1. Roberts, E.D.; Applied Polymer Symposium No.23; John Wiley & Sons Inc., New York, 1974; p.87-97.
2. Roberts, E.D.; A.C.S. Coatings and Plastics Preprints, 35, No.2, 281-6.
3. O'Keefe, T.W.; Vine, J., Handy, M.R.; Solid State Electronics, 1969, 12, 841-8.
4. Scott, J.P.; 6th Int. Conf. on Electron and Ion Beam Science and Technology, 1974, Edited by Bakish, Robert A. Electrochemical Soc. Inc., New York, N.Y. p.123-36.

5.   Beasley, J.P.;   4th Int. Conf. on Electron and Ion Beam
     Science and Technology, 1970.   Edited by Bakish, Robert A.
     Electrochemical Soc. Inc., New York, N.Y. p.515-23.
6.   Beasley, J.P., Squire, D.G.;   IEEE Trans. Elect. Devices,
     ED-22, 1975, (7), 376.
7.   Roberts, E.D.;   Vacuum, 1976, 26, (10/11), 459-67.
8.   Roberts, E.D.;   A.C.S. Coatings and Plastics Preprints,
     1980, 43, 231-5.

RECEIVED October 19, 1981.

Radiation Degradation and Film Solubility Rates of Poly(butene-1-sulfone)

LARRY STILLWAGON

Bell Laboratories, Murray Hill, NJ 07974

The solubility rate of a positive polymeric resist film exposed to radiation is higher than the solubility rate of the original unexposed film. The solubility rate increase depends upon the response of the developer to the physical and chemical changes which occur in irradiated film regions and has been presumed to result mainly from a molecular weight decrease caused by scisson of the polymer chains. Chemical structure of the polymer and film porosity may also be affected by irradiation and may therefore also influence the irradiated film solubility rate. Ouano (1) found that irradiated poly(methyl methacrylate), PMMA, films had higher solubility rates than unirradiated PMMA films of comparable molecular weight. The solubility rate difference was attributed to a higher solvent penetration rate into the irradiated films. These films were postulated to be more porous than unirradiated films because volatile matter, created during irradiation, was lost from the films.

Poly(butene-1-sulfone), PBS, rapidly degrades when exposed to radiation and is a highly sensitive electron beam resist (2). PBS films are exposed to 8×10^{-7} coul/cm^2 at 10kV (about 4 Mrads) to form useful lithographic images. About 190μ moles of gas, mostly butene and sulfur dioxide, are evolved when 1 gram of PBS is exposed to a 4 Mrad gamma-ray dose (3). The total weight of gas evolved per cm^2 from a 1μm thick PBS film exposed to 4 Mrads can be calculated to be about 1μgram. It has been calculated that about 0.2μgrams of gas is evolved per cm^2 of a 1μm thick PMMA film exposed to 1.4×10^{-5} coul/cm^2 at 2.5kV (>100 Mrads) (1,4). It was concluded from the above gas yield calculations that irradiated PBS films should have a microporous structure similar to that postulated for irradiated PMMA films.

The purpose of this paper was to establish a relationship between unirradiated PBS film solubility rates and molecular weight and then compare irradiated PBS film solubility rates to the solubility rates predicted by this relationship. Contributions to the solubility rate of irradiated PBS films from radiation-induced changes other than molecular weight decrease were assessed from such a comparison. Radiation G-values for PBS were determined as a first step in the comparison process.

0097-6156/82/0184-0019$05.00/0
© 1982 American Chemical Society

Experimental

Sample Preparation. PBS-MP20, obtained in powder form from Mead Chemical Co., was the primary source of the samples used in this study. A portion of the MP20 powder was dissolved in 2-methoxyethyl acetate (Mead Chemical) and films were spin-coated on silicon wafers. Films were baked at 120°C for 1 hour prior to irradiation. MP20 samples, both in powder and film form, were irradiated under vacuum at 30°C in a ^{60}Co source. Films were spun from methoxyethyl acetate solutions of the irradiated powders. These films were also baked at 120°C for 1 hour. The above procedure created two types of degraded MP20 samples: (1) those irradiated in film form and (2) those irradiated in powder form. The first type of films were designated as IF-type films. Films spun from solutions of the irradiated powders were designated as IP-type films.

Three PBS samples of lower molecular weight than MP20 were prepared by free-radical polymerization. Films of these samples were baked as above and were designated as U-type films. These samples were not irradiated in powder or film form.

Molecular Weight Measurements. Weight average molecular weights, $\overline{M}_w$, were measured using a Chromatix KMX6 low angle light scattering photometer. Number average molecular weights, $\overline{M}_n$, were measured using a Wescan Model 231 Recording Membrane Osmometer. Ethyl acetate was used as a solvent for the polymer in both techniques.

Solubility Rate Measurements. Solubility rates, S_R, were determined by measuring PBS film thickness as a function of development time. PBS films coated on silicon wafers were broken into several pieces. Each piece was dipped into n-butyl acetate, BuAc, and the development time measured. The film was rinsed in isopropanol and baked at 120°C for 30 minutes to remove residual solvents. Film thickness was measured by interferometry. The temperature of the developer, BuAc, was controlled at 25°C ± 0.05°C and the developer was not stirred or agitated during the development process. Plots of film thickness vs. development time were linear for low $\overline{M}_w$ films. Films having $\overline{M}_w$ greater than 400,000 g/mole did not completely dissolve in BuAc.

Results and Discussion

The radiation G values, G(s) and G(x), of a polymer can be determined by measuring molecular weight changes which occur during irradiation and applying equations 1 and 2 (5,6).

$$\overline{M}_w^{-1}(R) = \overline{M}_w^{-1}(0) + \left[\frac{G(s)-4G(x)}{1.93 \times 10^6}\right] R \qquad (1)$$

$$\overline{M}_n^{-1}(R) = \overline{M}_n^{-1}(0) + \left[\frac{G(s)-G(x)}{9.65 \times 10^5}\right] R \qquad (2)$$

G(s) and G(x) are defined as the number of scissons and the number of crosslinks produced when the polymer absorbs 100 eV. of energy and R is the radiation dose expressed in Mrads. Table I lists values of $\overline{M}_w$ and $\overline{M}_n$ measured for irradiated MP20 samples. A plot of $\overline{M}_w^{-1}$ and $\overline{M}_n^{-1}$ vs. radiation dose is shown in Figure 1. The solid lines in Figure 1 are linear least-squares fits to the data in Table I. Values for G(s) and G(x) were determined from the slopes of these lines to be 6.8 ± 0.9 and 0.3 ± 0.3, respectively.

TABLE I

Molecular Weights of Irradiated MP20 Samples

Dose (Mrad)	$\overline{M}_w \times 10^{-5}$ (g/mole)	$\overline{M}_n \times 10^{-5}$ (g/mole)
0.0	10.1	3.20
0.25	6.09	2.06
0.26	-	2.10
0.50	4.04	1.47
0.97	-	1.09
1.18	2.22	0.88
1.99	1.47	-
2.85	1.16	-
5.43	0.60	-

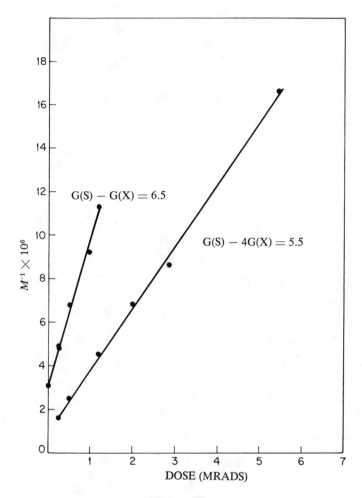

Figure 1.   Plot of $\overline{M}_w^{-1}$ and $\overline{M}_n^{-1}$ vs. radiation dose.

Table II lists measured solubility rates of three types of PBS films in the BuAc developer at 25°C. IP designates films spun from irradiated MP20 powders. The samples designated by IF are films of MP20 which were irradiated in film form. The third type, U-type, designates films spun from unirradiated PBS polymers other than MP20. It was assumed that a MP20 film on a silicon wafer (IF-type) would receive the same absorbed dose as MP20 powder if both were exposed to gamma radiation for the same period of time, i.e., the effect of back-scattered electrons created in the silicon wafer was ignored. A film spun from an irradiated powder, IP-type, and an IF-type film should therefore have the same $\overline{M}_w$ and chemical structure if the powder and film were exposed to irradiation for the same period of time. IF-type films were found to have higher solubility rates than IP-type films having a similar $\overline{M}_w$. A log S_R vs. log $\overline{M}_w$ plot of the data in Table II is shown in Figure 2. Uebereitter ($\underline{7}$) found that such log-log plots were linear for other polymer-solvent systems. The solid lines in Figure 2 are linear least-squares fits to the data in Table II. The equations describing these lines are also shown in Figure 2. The solubility rate difference between two film types, IF and IP, having the same $\overline{M}_w$ must arise from a difference in solvent penetration rate into the two film types since the thermodynamic contribution to S_R, i.e., polymer-solvent interactions, should be the same for both film types. The increased solvent penetration rate into an irradiated PBS film is consistent with the argument that escaping gases created during irradiation produce voids in the film (see reference $\underline{1}$).

Table III lists the contribution to the overall irradiated film solubility rate from the enhanced solvent penetration resulting from radiation exposure. The values for $S_R(IF)$ and $S_R(IP)$ in Table III were calculated using the equations shown in Figure 2. The % contribution to the overall solubility rate, 100 $[S_R(IF)-S_R(IP)]/S_R(IF)$, shown in Table III increased as the radiation dose increased. This behavior is expected since the number and possibly the size of the voids created in the film would be expected to increase as the radiation dose increased.

More evidence for enhanced porosity in irradiated PBS films is provided by post-exposure annealing experiments. The solubility rate of an irradiated MP20 film which was annealed for 30 minutes at 120°C before dissolution was lower than the solubility rate of the unannealed film. Table IV shows the measured solubility rate loss and the expected solubility rate loss, $S_R(IF) - S_R(IP)$, if all the porosity created during irradiation were removed by annealing. The solubility rate of the annealed IF-type films are somewhat higher than IP-films of similar $\overline{M}_w$. Thus, the annealing process removes most but not all of the porosity created during irradiation.

The chemical structure of PBS also may be altered by exposure to radiation and such changes may contribute to the solubility rate difference between an exposed and an unexposed PBS film. U-type films were prepared from unirradiated powders while IP-type films were prepared from irradiated powders. Inspection of Table II or Figure 2 shows that the three U-type films have slightly larger solubility rates than IP-type films of comparable $\overline{M}_w$. The solubility rate differences between IP and U-type films are small relative to the differences between IP and IF type films. The solubility rate difference between a U and an IP film of comparable $\overline{M}_w$ must arise from chemical structural differences between irradiated and unirradiated powders. These radiation-induced changes may also be responsible for differences observed in the elution behavior between irradiated and unirradiated PBS samples in gel permeation chromatography experiments. Irradiated PBS samples yield abnormally broad elution curves while unirradiated samples elute normally ($\underline{3,8}$).

TABLE II

PBS Film Solubility Rates in BuAc at 25°C

Dose (Mrad)	$\overline{M}_w^{**} \times 10^{-5}$ (g/mole)	$S_R \times 10^{-3}$ (Å/min)
5.28 (IP)*	0.62	4.71, 4.36
3.30 (IP)	0.96	4.06
2.01 (IP)	1.49	3.73
1.06 (IP)	2.50	3.17
0.50 (IP)	4.17	2.99
5.28 (IF)*	0.62	8.24
3.30 (IF)	0.96	7.07
1.96 (IF)	1.52	5.02
1.14 (IF)	2.37	4.66
0.50 (IF)	4.17	3.87
0 (U)*	0.85	4.89
0 (U)	1.08	4.59
0 (U)	2.72	3.36

*   Explained in text.

** Calculated from G-values for irradiated samples.

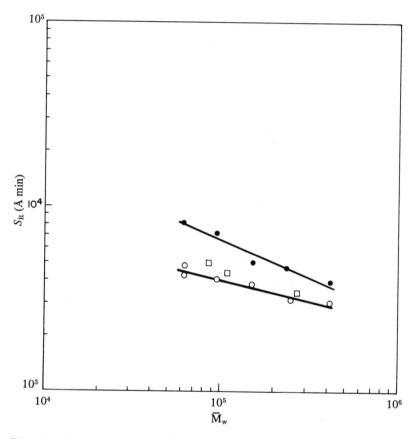

Figure 2.   Plot of solubility rate S_R in BuAc at 25°C vs. M_w.   Key: ●, $S_R = 7.6 \times 10^5 \overline{M}_w^{-0.41}$ (IF-TYPE); ○, $S_R = 5.6 \times 10^4 \overline{M}_w^{-0.23}$ (IP-TYPE); and □, U-TYPE.

TABLE III

Contribution to S_R of Irradiated PBS Film
From Enhanced Solvent Penetration*

Dose	$[S_R(IF) - S_R(IP)] \times 10^{-3}$	$100[S_R(IF) - S_R(IP)]/S_R(IF)$
(Mrad)	(Å/min)	
5.0	3.68	46%
3.0	2.72	41%
2.0	2.13	37%
1.0	1.39	30%
0.5	0.92	24%

* S_R calculated from equations in Figure 2.

TABLE IV

Post-Exposure Annealing Results

Dose	Measured S_R Loss $\times 10^{-3}$	Expected S_R Loss $\times 10^{-3}$
(Mrad)	(Å/min)	(Å/min)
4.90	2.96	3.40
3.05	1.90	2.53

Conclusions

The increased solubility rate of an irradiated PBS film can not be attributed solely to the decreased molecular weight of the irradiated PBS polymer. Part of the solubility rate increase is due to a higher solvent penetration rate into the irradiated film which is more porous than the original unexposed film. The above conclusions were reached by comparing the solubility rate of irradiated PBS films and films prepared from irradiated PBS powders (IF and IP films, respectively, see Table II or Figure 2). IF-type films were found to have higher solubility rates in BuAc than IP-type films having a similar molecular weight. The solubility rate difference between the two film-types increased as the radiation dose increased. Furthermore the solubility rate of an irradiated PBS film which was annealed before dissolution was lower than the solubility rate of the unannealed film. The results of this study indicate that radiation gas yields as well as G(s) values should be considered when choosing candidates for positive resists used with liquid developers.

Acknowledgements

The author acknowledges helpful discussions with M. J. Bowden and G. N. Taylor.

Literature Cited

1. Ouano, A. C., Polymer Eng. and Sci. 1978, 18, 306.
2. Bowden, M. J. and Thompson, L. F., Journal of Electrochem. Soc. 1973, 120, 1722.
3. Brown, J. R. and O'Donnell, J. H., Macromolecules 1971, 5, 109.
4. Hiraoka, H., IBM J. Res. 1977, 21, 121.
5. Saito, O., Poly Eng. and Sci. 1979, 19, 234.
6. Inokuti, M., J. Chem. Phys. 1963, 38, 1174.
7. Uebereitter, K., Chap. 7 in "Diffusion in Polymers", J. Crank and G. Park, Editors, Academic Press, New York, 1968.
8. Stillwagon, L. E., ACS Preprints, Org. Coatings and Plastics Chem. 1980, 43, 236.

RECEIVED October 19, 1981.

Poly(methyl methacrylate-co-3-oximino-2-butanone methacrylate-co-methacrylonitrile)

A Deep-UV Photoresist

E. REICHMANIS and C. W. WILKINS, JR.

Bell Laboratories, Murray Hill, NJ 07974

Photolithography is the most widely used technology for the production of integrated circuits today. Conventional projection equipment, which employs 350-450nm light, is diffraction limited to ~2μm resolution. The increasing complexity and miniaturization of integrated circuits however, requires the utilization of smaller features. One attractive approach to effect this goal, is the use of shorter wavelength (230-280 nm) light. It is expected that decreased diffraction will permit the formation of smaller features (≤1μm) (1—7). This however requires the development of new resist materials.

The ideal resist for deep UV lithography should possess good sensitivity to 230-280 nm radiation with little or no absorption at longer wavelengths to eliminate the difficult task of filtering the long wavelength radiation present in conventional sources. In addition, the resist should be capable of high resolution, have a reasonable exposure time, and be otherwise compatible with conventional microstructure fabrication processes.

Various materials have been examined for use as deep UV resists: poly(methyl methacrylate) (PMMA) (3), poly(methyl isopropenyl ketone) (PMIPK) (5,7), and the novolak-Meldrum's acid solution inhibition system (8). Each however has a problem related to sensitivity and/or resolution. While PMMA is insensitive to light of λ > 230 nm because of its weak absorption, its high resolution properties make it an attractive starting point for the design of a resist that will perform well in the 230-280 region. The photochemical properties of PMMA could be modified by the incorporation of a small percentage of photolabile groups so as to have both the desired sensitivity and base polymer properties.

One attractive possibility is the use of the α-keto-oxime chromophore. It has a strong absorption at ~220 nm whose tail, which extends to ~240-250 nm, would improve the absorption characteristics of PMMA. Also, the esters possess a N-O bond which is photochemically labile yet sufficiently thermally stable so as to be compatible with the various processing steps. The solution degradation of α-keto oximino methacrylate esters upon irradiation with light of λ365 nm has been reported by Delzenne (9), and we proceeded to investigate the solid state photodegradation of similar copolymers and their possible utility as deep UV photoresists.

0097-6156/82/0184-0029$05.00/0

EXPERIMENTAL

Polymer Preparation. Polymers were typically prepared in ethyl acetate solution at reflux temperature using benzoyl peroxide as initiator. They were isolated by two precipitations from ethyl acetate solution into methanol and were dried under vacuum. Molecular parameters of the polymers prepared are listed in Table II.
Materials:
Poly(methyl methacrylate) ((PMMA); Elvacite) is a high molecular weight polymer available from duPont which was used as a standard for measuring sensitivities.
Sensitivity:
Poly(methyl methacrylate-co-3-oximino-2-butanone methacrylate) (P(M-OM) and poly(methyl methacrylate-co-3-oximino-2-butanone-methacrylate-co-methacrylonitrile) (P(M-OM-CN)) were dissolved in methoxyethyl acetate (10% solution). Where appropriate, the specified amount of sensitizer was added to the solutions before coating onto a silicon substrate with a Headway Research spinner. Films were prebaked at 120°C for 60 min.

Photon sensitivities are measured relative to PMMA and were obtained by imaging a 1mm wide slit, illuminated by a 1000 watt Hg lamp focused through quartz condenser optics, onto the substrate for varying times. Exposure times were recorded as the time necessary to allow complete removal of the resist in the irradiated areas, with no thinning in the unexposed areas. The irradiated films were developed in methyl isobutyl ketone (MIBK)/2-propanol (7:3 v/v) for the copolymers and terpolymers, and MIBK for PMMA.

Resolution. The ultimate resolution capability of the materials was determined by electron beam exposure at 20kv and $5 \times 10^{-5} C/cm^2$.

RESULTS AND DISCUSSION

The incorporation of small percentages (<10%) of 3-oximino-2-butanone methacrylate (4) into poly(methyl methacrylate) (PMMA) (Scheme I) results in a four fold increase in polymer sensitivity in the range of 230-260 nm (10,11). Presumably, the moderately labile N-O bond is induced to cleave, leading to decarboxylation and main chain scission (Scheme II). The sensitivity is further enhanced by the addition of external sensitizers. Also, preliminary results indicated that terpolymerization with methacrylonitrile would effect an additional increase. These results complement those of Stillwagon (12) who had previously shown that copolymerization of methyl methacrylate with methacrylonitrile increased the polymer's sensitivity to electron beam irradiation. The mole fraction of the comonomers was kept low in order to insure retention of the high resolution properties of PMMA (3,4).

In an effort to improve PMMA's photosensitivity further, methyl methacrylate has been copolymerized with higher percentages of the α-keto-oxime methacrylate and terpolymerized with varying amounts of methacrylonitrile. The resulting effects on resist properties, e.g., sensitivity, contrast and resolution, and plasma resistance, are reported here. The terpolymers are up to 85 times more sensitive than PMMA, and retain its high resolution characteristics.

Sensitivity. Typical absorption spectra of the copolymer, poly(methyl methacrylate-co-3-oximino-2-butanone methacrylate) (P(M-OM)), are shown in Figure 1, and Table

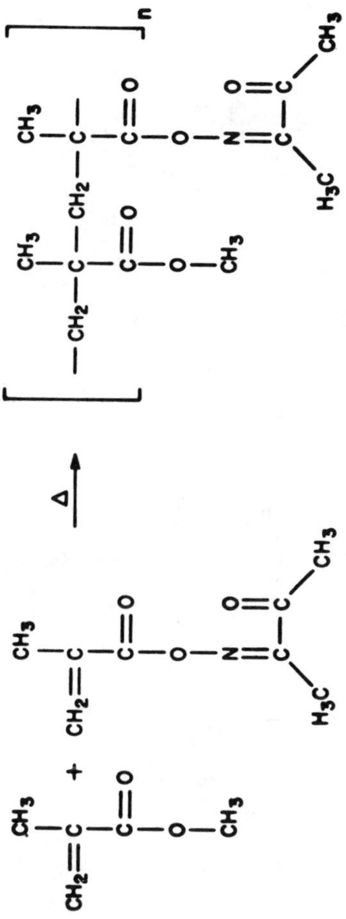

Scheme I.

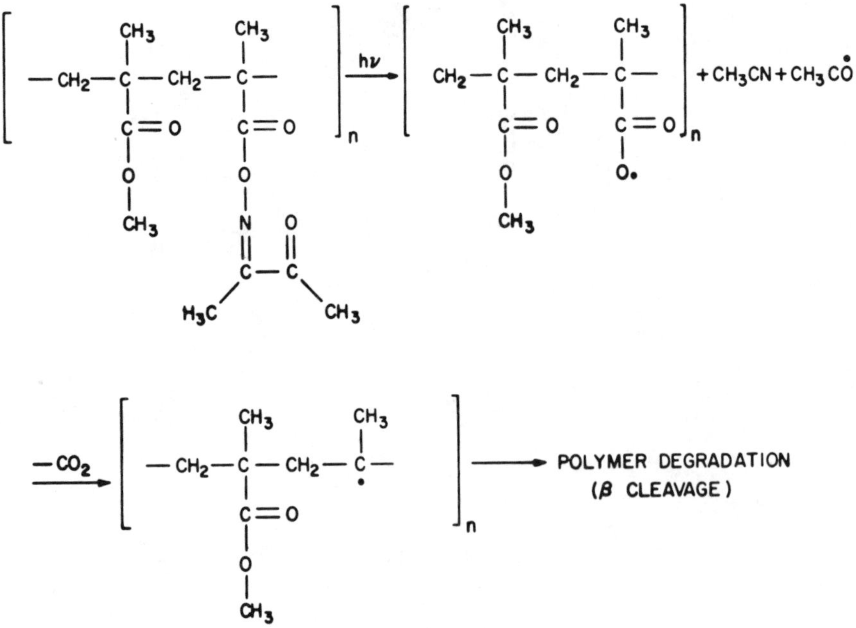

Scheme II.

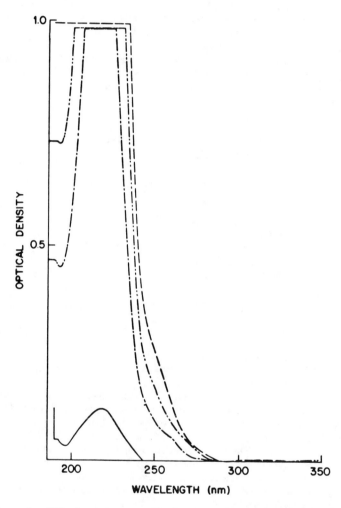

*Figure 1.   UV absorption spectra of 1-μm (nominal) films of PMMA (——),
P(M-OM) (94:6) (– · –), P(M-OM) (84:16) (– · · –), and P(M-OM) (63:37) (– – –).*

1 gives the optical density of nominal 1μm films of P(M-OM) as measured at the shoulder at 240 nm. As expected, this optical density increases with increasing oxime ester concentration. For instance, P(M-OM) (94:6) has an O.D. of ~0.2 at this wavelength, vs. 0.4 for P(M-OM) (84:16). While the O.D.'s of the (63:37) and (36:64) copolymers are probably too high to obtain a uniform exposure through the film, the materials containing up to ~20% OM should be acceptable. Examination of Table II, Group I reveals that increasing oxime ester concentration improves sensitivity. As measured relative to PMMA, incorporation of 16 mole% α-keto-oxime results in a 30-40 fold improvement in sensitivity on exposure to the full output of a mercury lamp. The monochromatic sensitivity of this material is ~$100mJ/cm^2$ at 240nm, and $500mj/cm^2$ at 265nm. An additional, though less dramatic improvement, is obtained by further increasing the α-keto-oxime mole fraction; P(M-OM) (63:37), is ~50 times more sensitive than PMMA. At this point however, the O.D. of the film is too high to allow a uniform exposure through the film.

Optical Density Data for P(M-OM)

TABLE I

Polymer	Composition	Optical Density[a] at 240nm
P(M-OM)	94:6	0.2
P(M-OM)	87:13	0.3
P(M-OM)	84:16	0.4
P(M-OM)	63:37	0.5
P(M-OM)	36:64	0.7

---

[a]Film thicknesses are nominally 1μm, but may vary by $\leq20\%$.

Incorporation of methacrylonitrile (Scheme III) effects a further enhancement in photosensitivity. Group II of Table II shows the effect of increasing the proportion of methacrylonitrile while keeping that of the oxime ester moiety constant. The value of 6 mole % oxime ester was chosen. The data indicate that poly(methyl methacrylate-co-3-oximino-2-butanone methacrylate-co-methacrylonitrile) (P(M-OM-CN)) exhibits an ~2.5 fold increase in sensitivity over the corresponding copolymer when the mole fraction of methacrylonitrile is 15-22%. On increasing this to 28%, the sensitivity begins to decrease. This latter effect is potentially due to the solubility characteristics of the polymer. Since the sensitivity of P(M-OM-CN) (86:6:8) is equivalent to that of P(M-OM) (94:6), it appears that a minimum amount of methacrylonitrile, between 8 and 15% in this instance, is required before any improvement in photosensitivity is observed.

The effect of varying the concentration of the 3-oximino-2-butanone methacrylate moiety was studied next. From the above results, it appears that the optimum nitrile concentration is roughly 15 to 22%. We chose to fix the methacrylonitrile mole fraction at the lower value in order to least perturb the solubility characteristics of the polymer.

P(M-OM) and P(M-OM-CN) Polymer Properties

TABLE II

Polymer	Composition	M_w (x10-5)	Mw/Mn	Contrast	Tg(°C)	Sensitivity Photo[a] 200-400nm	Electron beam[b] (x10^{-5}C/cm^2)
PMMA[c]	—	—	—	1.9	105	1	5
			GROUP I				
P(M-OM)	94:6	2.87	1.86	2.0	103	0.25	4
P(M-OM)	87:13	2.05	2.09	1.9	—	0.025	—
P(M-OM)	84:16	2.48	2.04	1.7	95	0.033	—
P(M-OM)	63:37	1.88	2.35	1.7	89	0.02	—
P(OM)	—	—	—	—	95[d]	0.01	—
			GROUP II				
P(M-OM-CN)	86:6:8	1.82	2.60	1.5	103	0.25	—
P(M-OM-CN)	79:6:15	3.41	2.66	1.5	99	0.1	5
P(M-OM-CN)	73:6:22	2.70	2.53	1.5	—	0.1	—
P(M-OM-CN)	66:6:28	1.40	2.30	1.4	—	0.2	—
			GROUP III				
P(M-OM-CN)	82:3:15	2.06	1.84	1	100	0.33	5
P(M-OM-CN)	79:6:15	3.41	2.66	1.5	99	0.1	5
P(M-OM-CN)	76:9:15	2.27	2.19	2	98	0.017	—
P(M-OM-CN)	73:12:15	2.97	1.92	1.8	95	0.017	—
P(M-OM-CN)	69:16:15	2.78	1.98	2	96	0.012	5

[a]Relative to PMMA in terms of exposure required

[b]at 20 kv

[c]duPont, Elvacite 2010, "high molecular weight"

[d]the material decomposes on heating

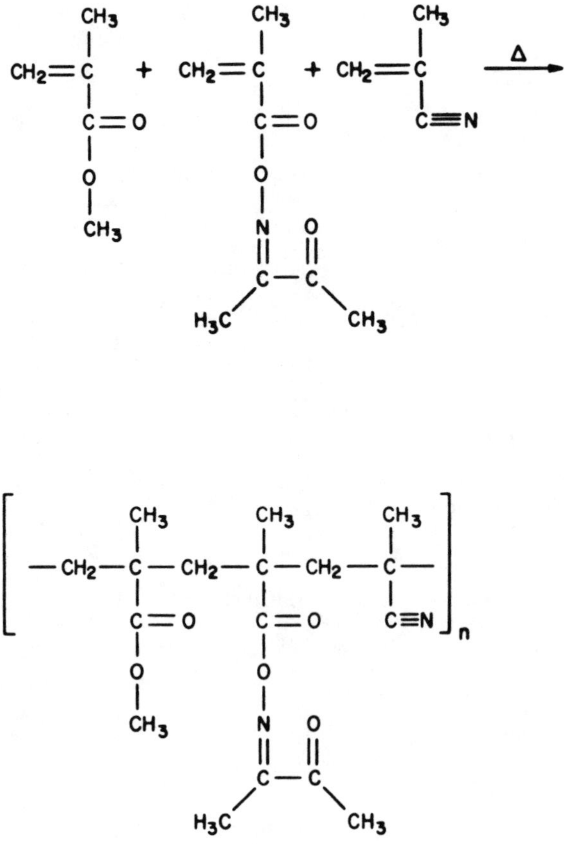

Scheme III.

Inspection of Table II, Group III indicates that the terpolymer sensitivity increases with increasing oxime ester concentration, a result which parallels that for the copolymers (Group I). P(M-OM-CN) (69:16:15), the most sensitive material prepared, is 85 times more sensitive than the parent PMMA. Note that the effect on the sensitivity of incorporation of methacrylonitrile is an additive one, *i.e.*, each terpolymer is roughly 2-3 times more sensitive than its corresponding copolymer.

Further improvements in sensitivity can be obtained by addition of external sensitizers which increase the amount of light absorbed in the film. The sensitizer must satisfy a number of criteria: i) its excited state energy must match (or be greater than) that of the keto-oxime chromophore; ii) it must transfer energy efficiently; iii) it must not absorb at longer wavelengths than the range desired; iv) it must be sufficiently non-volatile to remain in the polymer film during the initial baking; and v) it must be compatible with the polymer. One such material is p-t-butylbenzoic acid. Figure 2 shows a comparison of the absorption spectrum of P(M-OM-CN) (86:6:8) with that of the same material containing 15 wt% p-t-butylbenzoic acid. The effect on sensitivity of the addition of the benzoic acid to the "Group II" terpolymers is shown in Table III. An increase in photosensitivity by a factor of ~10, typical of all the systems in the present study, was found for these polymers when 15% sensitizer was used.

The Effect of p-t-Butylbenzoic Acid on Terpolymer Sensitivity

TABLE III

Polymer Composition	%Sens.[a]	Sensitivity[b]	Contrast	
P(M-OM-CN)	86:6:8	—	0.25	1.5
P(M-OM-CN)	86:6:8	15	0.017	1.6
P(M-OM-CN)	79:6:15	—	0.1	1.5
P(M-OM-CN)	79:6:15	15	0.008	1.5
P(M-OM-CN)	73:6:22	—	0.1	1.5
P(M-OM-CN)	73:6:22	15	0.012	1.6
P(M-OM-CN)	66:6:28	—	0.2	1.4
P(M-OM-CN)	66:6:28	15	0.033	1.6

[a] p-t-butylbenzoic acid
[b] Relative to PMMA, where PMMA = 1

While the mechanism for the photosensitization is not known, the results in Table IV indicate that it arises from the excited singlet rather than the triplet state. When P(M-OM) (94:6) was sensitized with the known triplet energy sensitizers, Michler's Ketone (4,4'bis(dimethylamino)benzophene, E_T = 61 Kcal/mole (13) or benzophenone (E_T = 68.5 Kcal/mole (13)), no effect on polymer photosensitivity was observed. However, in the presence of naphthalene-2-acetic acid, whose triplet energy is roughly the same as that of Michler's Ketone, a 3 fold increase in sensitivity obtains. Similarly, N-acetylcarbazole and p-t-butylbenzoic acid improve the polymers' photosensitivity. These sensitizers have reasonably long-lived excited singlet states that can transfer their energy to the oxime ester groups.

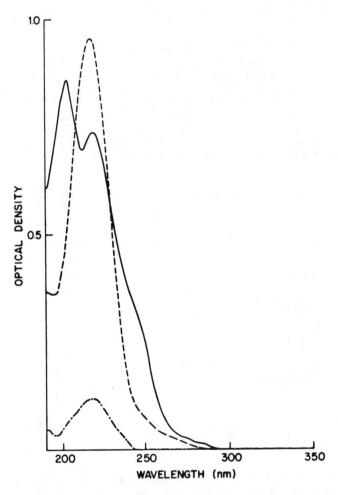

Figure 2.   UV absorption spectra of P(M-OM-CN) (8:0.6:1.4) (– – –) and P(M-OM-CN) (8:0.6:1.4)/15% p-t-butylbenzoic acid (———).

The Effect of Sensitizers on P(M-OM)

TABLE IV

Polymer	Sensitizer (%)	E_T (sens.)	Sensitivity[a]
P(M-OM)[b]	—	—	0.25
P(M-OM)	4,4'-Bis (dimethylamino) benzophenone (10)	61[c]	0.25
P(M-OM)	Naphthalene-2-acetic acid (10)	(61)[c,d]	0.08
P(M-OM)	Benzophenone (10)	68.5[c]	0.25
P(M-OM)	N-Acetylcarbazole (10)	68.3[e]	0.08
P(M-OM)	p-t-Butylbenzoic acid (10)	—	0.1

[a] Relative to PMMA
[b] P(M-OM) (94:6)
[c] Calvert and Pitts, "Photochemistry," John Wiley (1966)
[d] The value stated is that for parent naphthalene, which to a first approximation should have the same value
[e] M. Zander, *Ber. der Busengesellshaft*, 72, 1161 (1968).

The electron beam sensitivities of a random sample of co- and terpolymers were also determined, and were found to be essentially equal to that of PMMA. The value of $5 \times 10^{-5} C/cm^2$ was largely invariant with oxime ester concentration, and the presence of methacrylonitrile had no effect (13).

Contrast and Resolution. One of the parameters used to characterize the lithographic response of a material is contrast (γ), which is determined by taking the slope of the linear portion of the curve obtained on plotting the thickness of the relief image as a function of log (relative exposure dose). Generally, a high value of γ indicates that the material is capable of high resolution. Figure 3 depicts the contrast curves obtained for PMMA, P(M-OM) (87:13) and P(M-OM-CN) (69:16:15), whose contrast were found to be 1.9, 1.9, and 2.0, respectively. Typically, the contrast values were independent of α-keto-oxime or nitrile concentration. As may be seen by inspection of Table II, γ is affected by the dispersivity (Mw/Mn) of the polymer, the higher the dispersivity the lower the contrast, and thus the resolution would be expected to be adversely affected. These effects are generally seen in all resist systems (14). It is expected that the materials of contrast >1.8 should be capable of high resolution ($\leq 1 \mu m$).

The ultimate resolution capability of the terpolymers was determined by electron beam irradiation (20kv, $5 \times 10^{-5} C/cm^2$). Figure 4a depicts $0.5 \mu m$ and $1 \mu m$ lines and spaces printed in P(M-OM-CN) (69:16:15). Optical contact printing exposures (200-400 nm) are illustrated by Figure 4b which shows $0.75 \mu m$ lines and spaces, and Figure 4c which depicts profiles of $1 \mu m$ lines and spaces. These results are typical of all the materials examined.

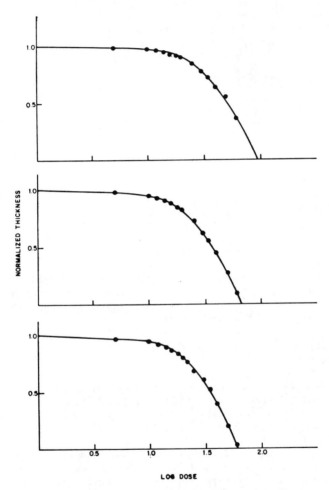

Figure 3.   Plot of log (dose) vs. normalized thickness for PMMA (top, $\gamma = 1.9$), P(M-OM) (87:13) (middle, $\gamma = 1.9$), and P(M-OM-CN) (69:16:15) (bottom, $\gamma = 2.0$).

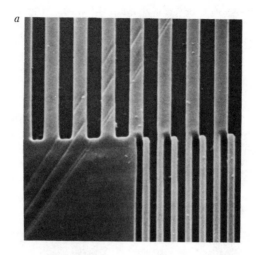

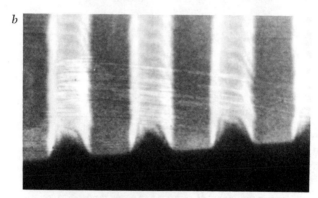

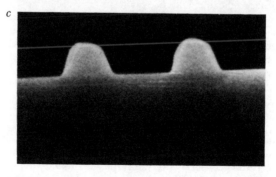

Figure 4.   SEM micrograph depicting: (a) 0.5- and 1.0-μm lines and spaces; (b) 0.75-μm lines and spaces; and (c) profiles of 1.0- μm lines and spaces printed in P(M-OM-CN) (69:16:15).

POLYMER MATERIALS FOR ELECTRONIC APPLICATIONS

Plasma Resistance. Another criterion that a resist must satisfy is that it be resistant to the processes used to transfer the relief image into the substrate. One of the more demanding processes is dry, or plasma etching.

Terpolymer Plasma Etch Data

TABLE V

Polymer	Composition	CF$_3$Cl[a]	Rate (Å/min.) C$_2$F$_6$-CF$_3$Cl(4:1)[b]	C$_2$F$_6$[c]	CF$_4$ - 8%O$_2$[d]
Polysilicon	—	800	1500	—	1000
SiO$_2$	—	—	—	1000	—
PMMA	—	350	>1000	350	>500
P(M-OM)	84:16	355	>1330	370	520
P(M-OM-CN)	69:16:15	310	1210	370	440
	73:12:15	—	—	—	490
	76:9:15	290	—	—	—
	82:3:15	—	>750	—	—
HPR-204[e]	—	100	720	100	350
AZ-111[f]	—	130	770	150	380
PIQ	—	200	970	210	420

---

[a]CF$_3$Cl gas at 200w and 0.35 torr
[b]C$_2$F$_6$-CF$_3$Cl (4:1) gas at 800w and 0.4 torr
[c]C$_2$F$_6$ gas at 800w and 0.4 torr
[d]CF$_4$-8%O$_2$ gas at 100w and 0.4 torr
[e]Hunt photoresist
[f]Shipley photoresist

A representative sample of terpolymers was exposed to a variety of etchants for polysilicon and silicon dioxide, and the results are given in Table V. The ratio of the etch rate of the substrate to the etch rate of the resist must be at least 2:1 for the resist to be a viable etch mask. Inspection of Table V, shows that the materials examined are unacceptable for only the C$_2$F$_6$ — CF$_3$Cl (4:1) plasma. The etch rates are comparable to those for PMMA; the α-keto-oxime exhibits essentially no effect on that rate and the nitrile affords a slight decrease in the plasma etch rate. The etch rates of some commercially available materials are shown for comparison.

CONCLUSION

The photosensitivity of PMMA is significantly enhanced by the incorporation of 10 to 40 mole% 3-oximino-2-butanone methacrylate. Terpolymerization with methacrylonitrile increases that sensitivity still further, P(M-OM-CN) (69:16:15) being 85 times more sensitive than PMMA on exposure to the full output of a 1000 watt mercury lamp. Upon addition of external sensitizers, this sensitivity may be increased by an additional factor of 2 to 3. The high resolution characteristics of PMMA have been retained and the polymers in question show good plasma resistance.

Acknowledgments

The authors wish to thank E. A. Chandross for many helpful discussions, M. Y. Hellman for the determination of polymer molecular weights, J. Frackoviak for the electron beam exposures and SEM micrographs, and A. C. Adams for the plasma etch data.

Literature Cited

1. Bowden, M. J.; Thompson, L. F. Solid State Technol., May 1979, pg. 72.
2. Appelbaum, J.; Bowden, M. J.; Chandross, E. A.; Feldman, M.; White, D. L. "Proceedings of the Kodak Microelectronics Seminar, Interface 75," Oct. 19-21, 1975.
3. Lin, B. J. J. Vac. Sci. Technol., 1975, 12, 1317.
4. Lin, B. J. IBM J. Res. Dev., 1976, 20, 213.
5. Nakane, J.; Tsumori, T.; Mifune, T. "Semiconductor International," 1979 (45).
6. Yamashita, Y.; Ogura, K.; Kunishi, M.; Kawazu, R.; Ohno, S.; Mizokami, T. "15th Symposium on Electron, Ion and Photon Beam Technology," Boston, Mass., May 1979.
7. Tsuda, M.; Oikawa, S.; Nakamura, Y.; Nakane, H. Photogr., Sci. Eng., 1979, 23, 290.
8. Wilson, C. G.; Clecak, N. J.; Grant, B. D.; Twieg, R. J. "Electrochemical Society Preprints," St. Louis 1980,p. 696.
9. Delzenne, G. A.; Laridon, U.; Peeters, H. Eur. Polym. J., 1970, 6, 933.
10. Wilkins, C. W. Jr.; Reichmanis, E.; Chandross, E. A. J. Electrochem. Soc., 1980, 127 (11), 2510.
11. Reichmanis, E., Wilkins, C. W. Jr.; Chandross, E. A. J. Electrochem. Soc., 1980, 127 (11), 2514.
12. Stillwagon, L. E.; Doerries, E. M.; Thompson, L. F.; Bowden, M. J. "ACS Division of Organic Coatings and Plastics Preprints," Chicago 1977, 37 (2), 38.
13. Calvert,J. G; Pitts,J.N. "Photochemistry," John Wiley and Sons Inc., 1966, pg. 298.
14. Thompson, L. F.; Kerwin, R. E. Annual Rev. Mat. Sci., 1976, 6, 267.

RECEIVED October 19, 1981.

4

Compositional Analysis of a Terpolymer Photoresist by Raman Spectroscopy

F. J. PURCELL and E. RUSSAVAGE
Spex Industries, Inc., Edison, NJ 08820

E. REICHMANIS and C. W. WILKINS, JR.
Bell Laboratories, Murray Hill, NJ 07974

The trend towards miniaturization in microstructure fabrication has created a demand for improved methods of production. The preceding paper (1) detailed one of the areas of research in this area, development of a deep UV-degradable photo-resist and presented a likely candidate, poly(methyl methacrylate-co-3-oximino-2-butanone methacrylate-co-methacrylonitrile) (P(M-OM-CN)).

While all of the results to date are encouraging, the exact composition of the terpolymer samples tested has been unknown. That information should be obtained in order to take full advantage of this resist. These polymers are especially resistant to the standard methods of analysis. Elemental analysis can be plagued by inaccuracies that arise from difficulties in determining low percentages of nitrogen and from residual solvent or monomers present in the polymers. UV spectrophotometry is useless because only the 3-oximino-2-butanone moiety yields a distinct UV spectrum. Pmr (proton magnetic resonance) spectroscopy has problems with overlapping absorptions. Only methyl methacrylate and the α-keto-oxime methacrylate have distinguishable resonance peaks, and only the methyl methacrylate can be integrated accurately. Thus, pmr can give no information about the methacrylonitrile and merely a rough estimate of the ratio for the other two. A possible answer to the problem may be found in C-13 nmr. However, this technique is rather time

0097-6156/82/0184-0045$05.00/0
© 1982 American Chemical Society

consuming due to the inherently weak carbonyl and nitrile signals caused by their long relaxation times. Infrared spectroscopy has problems with overlapping as well as weak absorptions.
Fortunately, each homopolymer does have a distinct Raman-active
band in addition to a band common to all components which can
serve as an internal standard. Thus, Raman spectroscopy provides a simple, nondestructive, and absolute method for the
determination of composition of P(M-OM-CN).

The Raman effect $(\underline{2},\underline{3})$ will be briefly described. When monochromatic radiation of frequency ν_o impinges on a sample, a small
portion of the light is scattered. Most of this is scattered
elastically; that is, it has the same frequency as the incident
light and is known as Rayleigh scattering. A much smaller percentage is also scattered inelastically with a frequency equal
to $\nu' = \nu o \pm \Delta\nu_M$, where $\Delta\nu_M$ is a Raman shift or Raman frequency.
For molecular systems, $\Delta\nu_M$ is generally associated with ro-
tational, vibrational or electronic transitions though only vibrational modes will be discussed in this paper. Transitions
from the ground state to a vibrationally excited state are
called Stokes lines and those originating in an excited state
are anti-Stokes bands. This is shown graphically in Figure 1.
Normally, only the Stokes spectrum is recorded because the intensity of the anti-Stokes spectrum is dependent on the popula-
tion of the excited states which follows the normal Boltzman
distribution. Thus, at room temperature, except for bands very
close to the exciting line corresponding to the lowest-lying
excited states, the intensity of anti-Stokes lines is greatly
reduced.

The intensity of Rayleigh scattering is on the order of
10^{-3} times the intensity of the incident light, and the Raman
intensities are at least 10^{-3} less than that of the Rayleigh
scatter. Thus, the Raman effect is obviously a weak phenomenon
which requires a high intensity monochromatic excitation source
(a laser) and a high dispersion spectrometer with excellent
stray-light characteristics to observe it.

The basic mechanism of the Raman effect is energy transfer.
An incident photon perturbs the molecule either giving up
energy or accepting energy from the molecule. Quantum mechanically, the incident photon is annihilated and a new photon of
lower or greater energy is created after interaction with the
molecule. Concomittant with this process is the creation or
destruction of a quantum of vibrational energy.

The selection rule for Raman spectroscopy requires a change
in the induced dipole moment or polarizability of the molecule,
and so it is a complementary technique to infrared which requires a change in the permanent dipole moment. For molecules
having a center of inversion, all Raman-active bands are infrared
inactive and <u>vice versa.</u> As the symmetry of the molecule is
lowered, the coincidences between Raman-active and infrared-

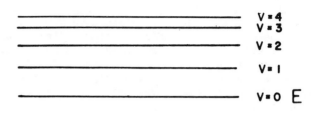
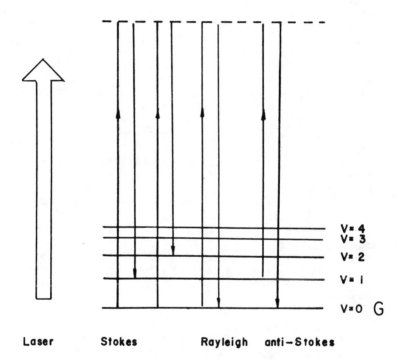

Figure 1. Energy level diagram depicting vibrational Stokes, Rayleigh, and anti-Stokes transitions.

active bands increase. However, because of the different selection rules, various functional groups have differing intensities in the two techniques. For example, water which is a
strong absorber in the ir is a very weak Raman scatterer. In
general, one can expect good Raman signals from large, deforma-
able groups such as S, I^-, unsaturated groups, $C = O$, and $C \equiv N$.
    Some of the information available from Raman spectroscopy
is qualitative; that is, it tells which functional groups are
present as determined from the characteristic group frequencies.
It may also be quantitative, supplying information as to the
amount of a particular substance present in a sample. This may
be on an absolute basis by means of an internal standard as in
the present case, or more normally, with the help of calibration curves. The information may also be structural. A
group-theoretical analysis of the data can provide a breakdown
of the number of vibrations according to symmetry groups for
various possible geometries of the molecule which, when compared
to the experimental data, may eliminate all but one of the possible geometries. Chain length can be calculated from the frequency of the longitudinal acoustic mode in polymer lamellae.
Additionally, conformational information of large molecules
may be obtained.
    Since the late 1960's a few papers have demonstrated com-
positional analysis of various polymer systems by Raman spectroscopy. For example, Boerio and Yuann (4) developed a method of
analysis for copolymers of glycidyl methacrylate with methyl
methacrylate and styrene. Sloane and Bramston-Cook (5) analyzed the terpolymer system poly(methyl methacrylate-co-
butadiene-co-styrene). The composition of copolymers of
styrene-ethylene dimethacrylate and styrene-divinylbenzene was
determined by Stokr et al (6). Finally, Water (7) demonstrated
that Raman spectroscopy could determine the amount of residual
monomer in poly(methyl methacrylate) to the 1% level. This was
subsequently lowered to less than 0.1% (8). In spite of its
many advantages, the potential of Raman spectroscopy for the
analysis of polymer systems has never been fully exploited.

Experimental

    All Raman spectra were recorded from samples in capillary
tubes with a SPEX RAMALOG Raman system consisting of a Model
1403 Double Monochromator, a Model 1459 Illuminator, and a
1460 LASERMATE tunable filter. The 514.5 nm line of a
Spectra-Physics Model 164-08 argon-ion laser supplied 0.16 to
0.2 W of power at the sample. The detection system consisted
of a cooled RCA C31034 GaAs photomultiplier tube and a SPEX
DPC2 digital photon-counting unit. The spectrometer was controlled, and all data manipulations were performed by the SPEX
SC32 SCAMP microprocessor data system. The spectra for the

compositional analysis were run with a 4 cm^{-1} spectral bandpass, and an integration time of 10 seconds (to optimize the S/N ratio). Only the peaks of interest were scanned in order to reduce the time of analysis to about 30 minutes per sample.

Results and Discussion

Figure 2 shows survey Raman spectra of the homopolymers, poly(methyl methacrylate)(PMMA), poly(3-oximino-2-butannone methacrylate)(POM), and poly(methacrylonitrile)(PMAN), and one terpolymer(P(M-OM-CN)) with a S/N ratio of about 10:1. Each of the polymers has a band specific to that polymer: 812 Δcm^{-1} (ν_s (C-O-C) for PMMA), 1622 Δcm^{-1} (ν_s(C=N) for POM), and 2237 Δcm^{-1}(ν_s(C≡N) for PMAN). Additionally, there is an asymmetric C-H bending mode at 1453 Δcm^{-1}, common to all three homopolymers, which serves as an internal standard. These bands are indicated by arrows in Figure 2. A broad fluorescence background is evident, but it can be reduced to acceptable levels by exposure to high laser power for 10-30 minutes, depending on the sample. Residual background fluorescence may be due to the oximino chromophore itself. Figure 3 depicts an example of actual data for a 75:15:10 terpolymer with a S/N ratio of about 50:1.

The compositional analysis was performed by first normalizing the spectra with respect to the integrated intensities of the internal standards of POM and PMAN to that found for PMMA. Once normalized, scaling factors for the three components were established by taking the ratio of normalized intensities of the 1622 (POM) and 2237 (PMAN) Δcm^{-1} bands to that of the 812 (PMMA) Δcm^{-1} band. The integrated areas of the scaled indicator bands for a terpolymer were summed to give the total area for all three components. The weight percent of each component was then obtained by dividing the area of the indicator band for each polymer by the total area. The same procedure was applied to the copolymers.

Table 1 gives a comparison of Raman and pmr results for a series of copolymers. In the pmr data of Figure 4, the CH$_2$ absorption of the polymer backbone at δ0.8 to 3.0 partially overlaps with the CH$_3$ doublet centered at δ2.4 and this reduces the accuracy of the integrated intensity of the ester moiety to no better than 25%. On the other hand, the accuracy of the Raman data is on the order of 5%, so the two techniques do agree within experimental error. The error associated with the Raman method could be reduced if calibration curves were employed. The weight percent feed and polymer compositions were converted to mole percent and reactivity ratios for MMA and OM were calculated by the Yezrielev, Brokhina and Riskin (YBR) method (9). The following equation, derived from the copolymer

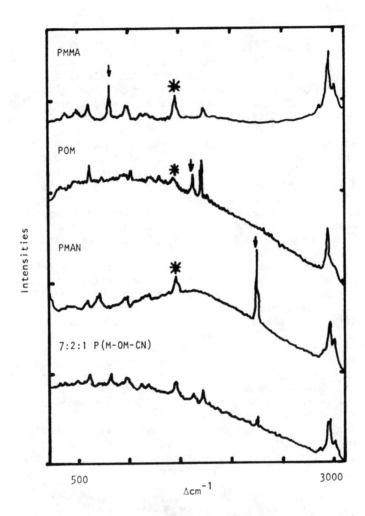

Figure 2.   Survey Raman spectra of the three homopolymers and a terpolymer. Key: 0.2-W power at 514.5 nm, 4-cm⁻¹ bandpass, 2-cm⁻¹ step size, 2-s integration time. The indicator bands are shown by arrows and the common internal standard band is denoted by an asterisk.

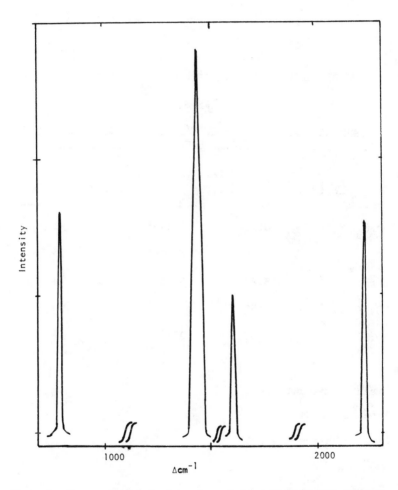

Figure 3. Raman data obtained for the analysis of 75:15:10 P(M-OM-CN) terpolymer. Key: 0.2-W power at 514.5 nm, 4-cm⁻¹ bandpass, 2-cm⁻¹ step size, 10-s integration time.

TABLE I

COMPOSITIONAL ANALYSIS OF P(M-OM) BY RAMAN AND PMR
SPECTROSCOPY

| Feed Ratio | | Copolymer Ratio | |
Mole %	Wt %	Raman (Wt %)	PMR (Wt %)
94:6	90:10	88:12[a]	91:9
90:10	85:15	82:18[b]	85:15
84:16	75:25	71:29[c]	74:26
63:37	50:50	50:50[a]	50:50
30:70	20:80	17:83[a]	–

[a]Mean of two runs

[b]Mean of three runs

[c]Mean of four runs

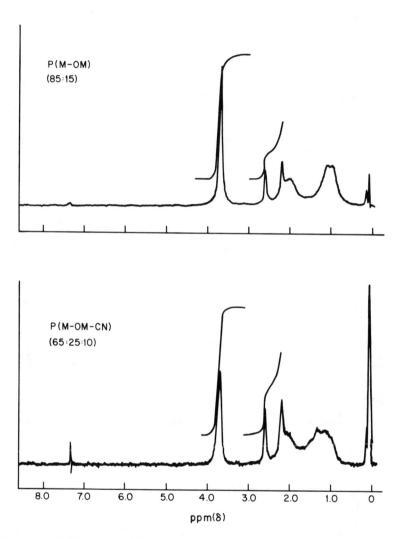

Figure 4. PMR spectra with integrations for CH₃ PMMA and POM absorptions for 85:15 P(M-OM) copolymer and 65:25:10 P(M-OM-CN) terpolymer. Data obtained from CDCl₃ solutions (50 mg/0.4 mL) on a Varian T-60 NMR spectrometer.

equation, was used:

$$\frac{f^{-\frac{1}{2}}-f^{\frac{1}{2}}}{f^{\frac{1}{2}}/F} = -\frac{F^2}{f} r_M + r_O$$

where $F = M_M/M_O$, M_M= mole fraction of MMA in the feed, M_O= mole fraction of OM in the feed; $f = m_M/m_O$, m_M= mole fraction of MMA in the copolymer, m_O= mole fraction of OM in the copolymer. Thus, r_M, the slope, is the reactivity ratio for MMA and r_O, the intercept, is the reactivity ratio for OM. The Raman data was employed to calculate f. The monomer reactivity ratios $r_M = 0.82\pm0.02$ and $r_O,= 0.94\pm0.08$ were obtained from the YBR plot (Figure 5) for the copolymerization of α-keto-oxime methacrylate and methyl methacrylate. Substituting these values into the copolymer equation (10), a drift of 1-6% is found for a conversion of 10-20%, which is insufficient to warrant compensation for varying the monomer feed composition. Figure 6 is a plot of the mole fraction of methyl methacrylate incorporated in the polymer (f_{MMA}) as a function of the mole fraction in the feed (F_{MMA}). The dashed line represents an ideal copolymerization.

Table II compares compositional analysis results by Raman spectroscopy with elemental analysis (C,H,N) for various mixtures of P(M-CN) copolymers; the agreement between the two methods is quite good and the results are consistent with published (11,12,13) reactivity ratios for this system. Because of the limited range of composition of the available samples of this copolymer, reactivity ratios were not calculated.

Table III lists compositional analysis results for several terpolymers. The Raman results are consistent with the reactivity ratios calculated for the P(M-OM) copolymer, and the published data for P(M-CN). These results indicate that the reactivity ratios for P(OM-CN) are of the same magnitude as those for P(M-OM) and P(M-CN).

In the pmr data for the terpolymer, overlap between the CH_3 absorption of the oxime ester and the backbone absorption is greater than in the copolymer pointed out in Figure 4. Thus, while the agreement between the Raman and pmr data for the terpolymer is not very good, (17-32% difference), it is completely within the experimental error of the pmr data. This large error and the fact that pmr can only distinguish two of the components of the terpolymer demonstrate that it is unsuited for compositional analysis of this system. Based on the agreement with published reactivity ratios and with the elemental analysis of the P(M-CN) copolymer, it is assumed that the Raman data are more accurate.

Since the polymer composition does not vary with polymerization, at least up to a conversion of 15-20%, it is unlikely that the similarity between the feed and product ratios results from the generation of a mixture of polymers of

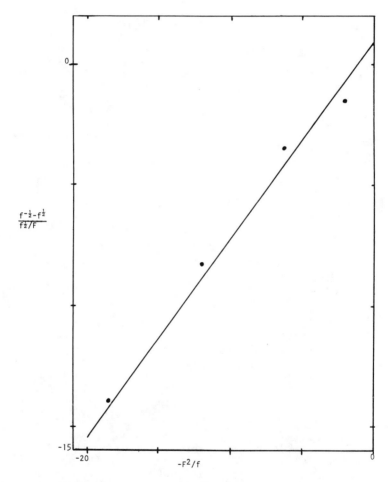

Figure 5. YBR plot for the copolymerization of 3-oximino-2-butanone methyl methacrylate and methyl methacrylate obtained from Raman data.

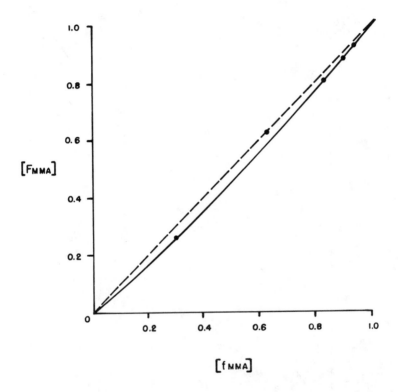

*Figure 6. Mole percent methyl methacrylate incorporated in poly(methyl)meth-
acrylate-co-3-oximino-2-butanone methacrylate) copolymers as a function of monomer feed composition determined by Raman spectroscopy. Key: — — —, ideality
and ———, experimental.*

TABLE II

COMPOSITIONAL ANALYSIS OF P(M-CN) BY RAMAN SPECTROSCOPY
AND ELEMENTAL ANALYSIS

| Feed Ratio | | | Copolymer Ratio |
Mole %	Wt%	Raman (Wt %)	Elemental Analysis (Wt %)
94:6	96:4	95:5[a]	96:4
88:12	92:8	92:8[b]	92:8
78:22	84:16	84:16[b]	85:15

[a]Mean of three runs

[b]Mean of two runs

TABLE III

COMPARISON OF COMPOSITIONAL ANALYSIS OF P(M-OM-CN)
BY RAMAN AND PMR SPECTROSCOPY

| Feed Ratio | | Terpolymer Ratio | |
Mole %	Wt %	Raman (Wt %)	PMR[a] Wt %
83:3:14	85:5:10	84:6:10	86:4
73:12:15	70:20:10	70:20:10	72:18
55:32:13	65:25:10	65:26:9	69:21
73:6:22	75:10:15	72:11:17	76:9

[a]Ratio of MMA to OMA

differing composition. Rather, the product is probably a polymer of uniform composition. Reported reactivity ratios for
various methacrylate esters(14), and for methacrylonitrile
(11, 12, 13)support the results obtained.

The limit of detection by Raman spectroscopy was 3-5
weight % for the oxime ester and methacrylonitrile for these
samples. The shorter time required to reduce background fluorescence in those samples filtered through activated charcoal
indicates that more careful sample preparation and purification
would lower this limit.

Conclusion

Knowledge of parameters such as reactivity ratios, is necessary for synthesis of polymer based resists, and an accurate
method of analysis should be useful in various areas associated
with resist development such as quality control. Raman spectroscopy provides a convenient, absolute, nondestructive method
for compositional analysis of polymer systems which, if an internal standard is present, does not require standards of known
composition or ancillary calibration curves. The accuracy, with
appropriate selection of experimental conditions such as slit
width and integration time, is limited only by the instru-
mentation.

Acknowledgment

We wish to thank L. E. Stillwagon for providing the P(M-CN)
polymer samples and elemental analysis results.

Literature Cited

1. Reichmanis, E.; Wilkins, Jr., C.W. preceding paper.
2. Long, D.A.; "Raman Spectroscopy;" McGraw-Hill International
    Book Company, New York, 1977; p1-132.
3. Szymanski, H.A., Ed.: "Raman Spectroscopy, Theory and
    Practice" Vol. 1; Plenum Press, New York, 1967; p1-44.
4. Boerio, F.J.; Yuann, J. K. J. Poly. Sci., 1973, 11, 1841.
5. Sloane, H.J.; Bramston-Cook R., Appl. Spectros. 1973,
    27, 217.
6. Stokr, J.; Schneider, B.; Frydrychova, A.; Coupek, J.
    J.Appl. Poly. Sci., 1979, 23, 3553.
7. Waters, D.N., "Proc. 5th Int. Conf. Raman Spectros.;"
    Hans Ferdinand Schulz Verlag, Freiburg Im Breisgau,
    1976; p 500.
8. Yu, N.T. Spex Speaker 1978, 23, 9.
9. Yezrielev, A.I.; Brokhina, E.L.; Roskin, Y.S. Vysokomol.
    Soedin., 1969, A11, 1670.

10. Billmeyer, Jr., F.W., "Textbook of Polymer Science" 2nd Ed.; John Wiley and Sons, New York; p 330.
11. Lewis, F.M.; Walling, C.; Cummings, W.; Briggs, E.R.; Wenisch, W.J. J. Am. Chem. Soc., 1948, 70, 1527.
12. Young, L.J., J. Poly. Sci. 1961, 54, 411.
13. Cameron, G.G.; Grant, D.H.; Grassie, N.; Lamb, J.E.; McNeill, I.E. J. Poly Sci., 1959, 36, 173.
14. Bandrup, J., Immergut, E.H., ed. "Polymer Handbook" Interscience, 1966; pII 195-198 and II 204-213.

RECEIVED October 19, 1981.

Effect of Composition on Resist Dry-Etching Susceptibility

Vinyl Polymers and Photoresists

J. N. HELBERT and M. A. SCHMIDT

Process Technology Laboratory, SRDL, Motorola, SG, Phoenix, AZ 85008

Plasma and reactive-ion etching and ion-milling etch rates for a group of vinyl polymer resists and photoresists have been determined and found to vary by over a factor of 20 as the vinyl side-group substituents were altered synthetically or the photoresist varied. Lower etch rates or better etch compatibilities are observed for vinyl polymer resists containing multiply-bonded and unsaturated side-groups, as well as for lithographically negative behaving systems. Generally, better etch compatibilities were observed for the photoresist systems although a couple of the vinyl systems did perform as well. The etch selectivities and trends, measured versus PMMA or SiO_2, are reasonably constant for the three etch techniques, except for the negative photoresists.

Polymer resists (1) serve as masking layers in the patterning of the dielectric and conducting layers encountered in integrated circuit (IC) fabrication. When IC device geometries were greater than 5 micrometers, the critical dimensions of these IC patterns could be controlled reasonably well—even though the wet chemical etching processes employed were purely isotropic. As IC geometries are shrunk from 3 to 1 micrometer and below, established wet isotropic techniques will have to be abandoned in favor of more anisotropic ones, in order to maintain etched geometry critical dimension control. As a result, the electronics industry has moved heavily into the area of dry-etching using plasma techniques (2) which are capable of achieving the desired etching anisotropy and dimensional-control. Obviously, polymer resists with good plasma etch resistance, or more generally, with "dry-process" compatibility, are in demand. This situation has evolved to the point that dry-etch resistance or compatibility has become the most important design criterion for new resists. It is this resist property that will determine resist polymer applications in the near future.

Etch resistance data for several polymer resist systems have been reported by Taylor and Wolf (3) and Moreau (4). Some of the results are tabulated in Table I. While Taylor and Wolf (3) have

0097-6156/82/0184-0061$05.00/0

Table I. CF_4/O_2 versus O_2 Plasma Etch (PE) results.

System	CF_4/O_2 PE Rate Ratio vs SiO_2	O_2 PE Rate Ratio vs. SiO_2[a]
Poly(N-vinyl carbazole)	–	0.25
Poly(styrene)	0.1	0.43
AZ 1350	0.3[b]	–
Poly(vinylidene flouride)	0.5[b]	0.83
Poly(methyl methacrylate)	1.0	1.0

[a] Data of G. Taylor and T. Wolf, reference 3.
[b] Data of W. Moreau, reference 4.

focused upon measuring polymer resist etching rates (or constants) for O_2 etched exposed polymers, Moreau (4) has focused upon determining resist etch resistances to the CF_4/O_2 plasma system, which is used in actual dielectric layer etching processes as opposed to resist ashing removal applications for the O_2 system. It is easy to see that the observed etch rates tabulated are influenced significantly by the resist polymer composition. The aromatic polymers at the top of Table I are clearly most resistant to etching. Curiously, the etch resistance trend is maintained regardless of plasma type or plasma reactive species, and depends more upon the resist polymer composition or structural formula.

In this work, we were particularly interested in expanding Moreau's limited list of CF_4/O_2 etch tested samples with a group of polymer resists, where the composition is known and side chain groups have been systematically altered to determine chemical moiety effects upon plasma etch resistance. We were also interested in determining the effect upon the resist polymer etch selectivities and selectivity trends caused by dry-etching with more anisotropic and potentially damaging systems, namely reactive-ion etching and ion-milling.

All three of the chosen dry etching techniques are capable of anisotropic etching; the specific CF_4/O_2 plasma etching (PE) system employed happens to be isotropic, but it is the most representative of the existing halogen-based plasma etching techniques. Plasma and reactive-ion etching (RIE) are governed primarily by halogen free radical chemistry (i.e., the halogen radicals or other halocarbon moieties produced in the plasma react with the samples to produce volatile halogen compounds). Reative-ion etching conditions differ from those of PE as RIE is carried out at lower pressures and higher electrical power; thus, RIE etching is aided and assisted by ion bombardment. RIE and ion-milling (IM) are also more anisotropic, due to the parallel nature of the electrodes employed and reduced gas pressures. Ion milling is carried out at lower pressures and the etching species is usually Ar^+, not halogen free radicals or fluorocarbon ions. IM etching is a physical process and is governed less by chemical reactivity, unless the ion employed is also reactive.

We report here plasma etch rate data for a series of vinyl resist polymers with a wide range of side chain substituents. The results of this study are valuable because they provide, when combined with other radiation chemical test data, improved design criteria for making improved high performance radiation resists. Structural fomulae and chemical nomenclature plus acronyms for the vinyl polymer systems studied are compiled below:

X,Y

CH_3, CO_2CH_3 - poly(methyl methacrylate) (PMMA)
Cl, CO_2CH_3 - poly(methyl alpha-chloroacrylate) (PMCA)
F, CO_2CH_3 - poly(methyl alpha-fluoroacrylate) (PMFA)

$(CH_2-C)_n$ with X top and Y bottom

X,Y

H, C_6H_5Cl - poly(chlorostyrene) (PCS)
H, C_6H_5 - poly(styrene) (PS)
Cl, CN - poly(MCA-co-methacrylonitrile)(P(MCA-co-MCN))

$X=CH_3,Y$

CO_2CH_3 - poly(methyl methacrylate) (PMMA)
$CO_2CH_2CH_3$ - poly(ethyl methacrylate) (PEMA)
$CO_2CH_2CH(CH_3)_2$ - poly(isobutyl methacrylate) (PIBM)
$CO_2C(CH_3)_3$ - poly(tert-butyl methacrylate) (PTBM)
$CO_2CH_2CCl_3$ - poly(trichloroethyl methacrylate) (PTCEM)
$CO_2CH_2CF_2$ - poly(trifluoroethyl methacrylate) (PTFEM)
$CO_2CH(CF_3)_2$ - poly(hexafluoroisopropyl methacrylate)(PHFIM)
CN - poly(methacrylonitrile)(PMCN)

These vinyl systems were chosen also because they function as high-resolution electron beam resists and deep UV resists at $\lambda < 300$ nm.

PE, RIE and IM resistances for an extensive list of commercial photoresists are included as well for comparison with the vinyl systems and amongst themselves. Although the exact comosition of these systems is not public information, the generic type of base resin or polymer binder is generally known. In addition, the photoactive components are all known to be aromatic azides or azo-compounds.

Experimental

The polymer resists used in this study were either synthesized in-house or obtained from Aldrich Chemicals or Polyscience, Inc. Photoresist samples were obtained from KTI Chemicals or the manufacturer. The polymers were dissolved in suitable solvents and spin coated onto oxidized Si wafers or Cr-coated glass test substrates. The polymer film thicknesses were measured either by a Taylor-Hobson proficorder or Tencor Alpha-step.

PE etch rate measurements were made in either a IPC 4005 or Tegal 421 system. To determine the plasma etch rate of the polymer

resist films, they were exposed to the etch for a specified time
interval, then the original resist step and a new step are measured with the profiling tools above. The change in the resist is
obtained by subtracting the resist step made after etch from the
original resist thickness. The silicon dioxide loss is measured
by stepping down from the post etch step to the original resist
step, where the oxide had been previously exposed to the plasma
etch. The selectivity vs SiO_2 is simply the ratio of resist loss
to oxide loss.

Resist reactive-ion etching (RIE) was performed with a totally modified Tegal Model 400 plasma reactor. Ion-milling (IM) was
accomplished with a Veeco three inch system. All resist RIE and
IM etch rates are measured versus the rate of SiO_2 and PMMA as
outlined above.

Because PMMA functions as a high resolution e-beam, x-ray and
deep UV resist (1), it was also used as a reference for all the
etch rates reported. The relative etch rate of each polymer resist was measured with PMMA under equivalent conditions, taking
the du Pont Elvacite 2041 rate to be 100 Å/min; the selectivity
of PMMA to oxide is 1.0. As the molecular weight of PMMA is
changed from 33K to 950K, the selectivity changes from 0.9 - 1.2.
Thus, as for thermal degradation the etch selectivity or etch
stability decreases with decreasing molecular weight.

Results and Discussion

To study the steric effect of the ester group (i.e., Y group)
upon polymer plasma etch resistance, the etch rates of PIBM, PTBM,
PEMA, and PCHM poly(methacrylates) were determined versus PMMA.
While PIBM and PTBM are observed to plasma etch a little slower
than PMMA, PEMA is surprisingly less resistant and PCHM more
resistant (see Table II). Since plasma, thermal and radiation
degradation processes are known to proceed via free radical inter-
mediates, the results of the Table are correlated with the latter
two properties. Methacrylate polymers containing alkyl groups
with β-hydrogens, such as ethyl, i-butyl, tert-butyl and cylohexyl,
all thermally degrade to ester group olefin. PMMA, on the other
hand, degrades directly to the MMA monomer. By analogy to the
thermal process, PMMA would be predicted to be less resistant
towards PE than the other poly(methacrylates), which is observed
experimentally. PEMA is the exception here, but PEMA is also the
only polymer exhibiting higher radiation degradation susceptibility towards ionizing radiation degradation. It appears that the
smaller the polymer ester alkyl group, the greater the ease of (1)
thermal degradation via depolymerization, (2) radiation degradation via decarboxylation, and (3) plasma degradation via plasma
species abstraction reactions. The PE rates observed cannot be
accounted for solely on the basis of thermal stability, because
the less susceptible polymer resists also have the lowest T_g
values (see PIBM and PCHM values of Table II).

Table II.   Relative etch rate ratios for ester X-substituted methacrylates, $\{CH_2-C(CH_3)CO_2X\}$.

	PE[a] Rate Ratio vs SiO$_2$	RIE[b] Rate Ratio vs SiO$_2$	G_s	G_x	T_g
$-CH_3$	0.9 - 1.2	1.2(RD)[c]	1.3	0	105
$-CH_2CH_3$	1.7	-	1.7	0	65
$-CH(CH_3)_2$	0.9	-	1.1	0	53
$-C(CH_3)_3$	0.8	-	1.3	0	118
$-Cyclo-C_6H_{11}$	0.4	-	$G_s-G_x=0.4$	0	66
$-CH(CF_3)_2$	2.2	-	2.6	0	80
$-CH_2CF_3$	2.2	RD[c]	2.3	0	69
$-CH_2CCl_3$	2.3	-	1.7;2.5	0.06	123
(1:1 MMA cop)	-	1.9	2.5	0.04	-
$-CH_2CF_2CFHCF_3$ (FBM-110)	1.4	RD[c]	-	-	50

[a] PE conditions: CF$_4$/8% O$_2$, 0.7 Torr, 100 Watts, on a Tegal 421 or CF$_4$/4% O$_2$, 0.5 Torr, 150 Watts, on a IPC 4005
[b] RIE conditions: CHF$_3$, 0.13 Torr, 350 Watts, on a Modified Tegal 400A
[c] RD = resist surface severly degraded

Table II also includes the PE rates for several poly(methacrylates) where the ester alkyl group is halogenated. Like PEMA the trichloro and fluoroethyl methacrylates are less resistant than PMMA or the oxide reference. In addition, the effect of halogenation has increased the PE degradation susceptibility by 30%. This effect is attributed to either ester alkyl group β-hydrogen sensitization towards radical scavenging created by the presence of the electronegative halogens on the adjacent carbon or to a degradation via a decarboxylation mechanism and sensitization. The larger observed G_s values are supportive of the latter explanation. For the poly(methacrylates) of Table II, there is a correlation between etch rate and G_s (see upper line in Figure 1). It must be emphasized here, however, that this is a select group of structurally-similar poly(methacrylates) with similar radiation and thermal characteristics. As the Figure illustrates, the general correlation with G_s is much weaker.

PMFA, a fluorinated poly(acrylate) and non-aromatic negative-behaving resist, possesses better PE and RIE resistance than PMMA (see Table III). The Table also includes results for PMCA, another polymer with an α-halogen. These results are interesting, because both polymers have a high T_g value (i.e., >130°C), and therefore, are more thermally stable than the polymers of Table II. Surprisingly, PMFA etches slower and is more resistant, while PMCA is significantly less resistant. The α-chlorine is known to enhance the radiation degradation susceptibility for PMCA vs PMMA as verified by a higher G_s value. (5) Since the C-Cl bond is readily cleaved, (5,6) it is easy to envision enhanced PE and RIE

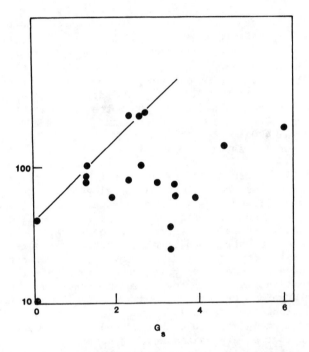

Figure 1. PE rate in Å/min vs. G_s for the vinyl polymer resist systems. The etching conditions are found in Table II.

Table III.   PE and RIE results for alpha-substituted polyacrylates $\{CH_2-C(X)CO_2CH_3\}$.

X		PE[a] Rate ($\overset{o}{A}$/min)	PE Rate Ratio vs SiO$_2$	RIE[b] Rate Ratio vs SiO$_2$	G_s	G_x
$-CH_3$	(PMMA)	93–110	0.9–1.2	1.2	1.3	0
$-Cl$	(PMCA)	186	1.8	1.2	6.0	0.9
$-F$	(PMFA)	40	0.4	0.6	0	1.0

[a]   See PE conditions of Table II
[b]   See RIE conditions of Table II

degradation susceptibility for PMCA, via absorption of $h\nu$ by the polymer and cleavage of the C–Cl bond in the plasma.   The C–F bond in PMFA, on the other hand, is not easily cleaved,[7] hence, PMFA would be less PE and RIE degradable.   The G_s values of Table III and the data of the cited references support this rationalization. PMFA is also similar in structure to that of poly(vinylidene fluoride), $\{CH_2-CF_2\}$, which has been reported in references 3 and 4 to be highly resistant to PE degradation (see also Table I). PVDF has not been investigated as a resist due to poor solubility, but it is known to predominantly crosslink when irradiated by ionizing radiation like PMFA. [8]   The etch resistances of PVDF and PMFA are most certainly governed by the strong C–F side chain bond(s).

The polymer resist exhibiting the lowest PE rate or highest etch resistance versus PMMA or oxide is poly(styrene) (see Table IV).   This system, like the others of Table IV, is representative of a vinyl polymer with general structural formula of $\{CH_2-CXY\}$. Poly(chlorostyrene), a chlorinated derivative of the aromatic poly(styrene), exhibits equal resistance towards all three dry etch processes.   Here halogenation has not enhanced the etch rate or reduced the resistance as seen before for PTECM, PTFEM, and PMCA nonaromatic systems.   Therefore, the aromatic side group must

Table IV: PE, RIE and IM results for di-substituted vinyl polymers $\{CH_2-CXY\}$.

$\underline{X}, \underline{Y}$	PE[a] Etch Rate Ratio vs SiO$_2$	RIE[b] Etch Rate Ratio vs SiO$_2$	Ion-milling[c] Rate Ratio vs SiO$_2$	T_g	G_s	G_x
$-CH_3$, $-CO_2CH_3$	0.9–1.2	1.2[d]	1.2	105	1.3	0
$-CH_3$, $-CN$	0.3	0.4	0.9	120	3.3	0
$-H, -C_6H_5$	0.1	0.1	0.7	100	0	0.05
$-H, -C_6H_4Cl$	0.1	0.3	0.8	–	–	–

[a]   See PE conditions of Table II
[b]   See RIE conditions of Table II
[c]   IM conditions; Ar, 0.9×10^{-4} torr, 600 volts, 15° angle, on Veeco 3" system
[d]   RD = resist surface severely degraded

dominate and cause the reduced PE degradation susceptibility of these vinyl polymers. These results are consistent with those listed in Table I taken from the literature. It also is evident that negative-behaving e-beam resists, like PS, PCS, PMFA and the others are generally more PE and RIE resistant than the other vinyl polymers. This may be attributed to the fact that when the α-hydrogens are abstracted by radical species, a crosslinking site is created and not an unstable degradation intermediate.

Of the polymer resists with structural formula $+CH_2-CXY+$ from Table IV or of the group of positive behaving e-beam vinyl resist polymers, PMCN is the most resistant (see Table IV and V). This etch resistance is attributed to the CN side chain group, which is a strongly bonded group that cannot be readily cleaved from the polymer backbone by radiation. It is notable that PMCN does not thermally degrade to monomer like other vinyl polymers, except at temperatures greater than $270^{\circ}C$; below $270^{\circ}C$ the polymer is more thermally stable than the other vinyls. Thus, the same stabilizing reaction, as that which occurs thermally over the temperature range of $100-270^{\circ}C$ to produce a ladder-like polymer containing $+N=\overset{.}{C}-N=\overset{.}{C}+$ units parallel to the main chain, may be occurring during dry-etching.

Table V: Electron beam positive (top) and negative (bottom) resist etching rate ratios for three sets of etching conditions.

Resist	CF$_4$/O$_2$PE[a] Rate Ratio vs SiO$_2$	RIE[a] Rate Ratio vs SiO$_2$	Ion Milling[a] Rate vs SiO$_2$	E-Beam Q,C/cm^2
PBS	∿9 – 10	–	–	0.7 – 1 x 10^{-6}
PHFIM	2.4	RD	–	1 x 10^{-5}
PTFEM	2.2	RD	–	1 x 10^{-5}
PMCN	0.3	0.4	0.9	2 x 10^{-5}
PMMA	0.9 – 1.2	1.2[a]	1.2	5 x 15^{-5}
AZ2400	0.5	0.6	0.5	1 x 10^{-4}
PC 129	0.2	0.2	0.5	2 x 10^{-4}
PS	0.1	0.1	0.7	1.5 x 10^{-5}
PMFA	0.4	0.6	–	2 x 10^{-5}
COP (KTI)	0.7	0.2	1.6	0.8 – 1 x 10^{-6}
SEL-N	0.6	0.6	1.2	0.8 x 10^{-6}
OEBR-100	0.6	0.6	1.1	–

[a] See etching conditions of Table IV

Consistent with the PMCN homopolymer results above, the MCN copolymers of Table VI exhibit PE etch rates intermediate to those of the two respective homopolymer values. When MCN is copolymerized with MCA, the resulting copolymers etch faster than PMCN and slower than PMCA (see Table VI). The etch rate is approximately linear with mole % MCA content; this data is plotted in Figure 2.

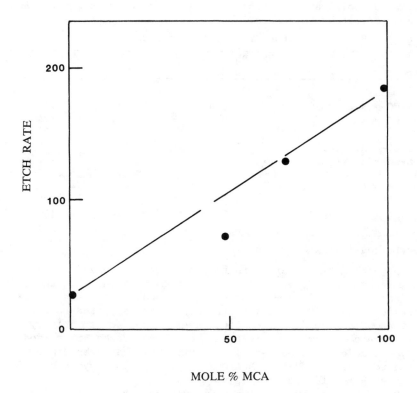

Figure 2.   PE rate in Å/min vs. mole % MCA for the MCN/MCA vinyl copolymer.   The etching conditions are found in Table II.

When MCA is copolymerized with MMA, the resulting copolymer etches
faster than PMMA, consistent with the PMCA and MCN/MCA results.
Incorporation of MCA, a monomer with α-chlorine, has the effect of
decreasing plasma etch resistance, as is observed for the MCA
homopolymer. Sensitization by chlorine appears to be general (see
Tables I, III, and VI), except for the case where the chlorine is
incorporated onto the aromatic side chain group. This effect was
observed previously by Taylor and Wolf for other polymer systems.
(3) The C-Cl bond strength is lower than that for C-H and C-F,
and there is much evidence that this bond can be easily cleaved,
even in the solid state. (5,6) This weaker side chain group leads
to lower dry-etch resistance.

Table VI: Plasma etch rate ratios for copolymers.

System		PE[a] Rate Ratio vs SiO$_2$	T$_g$
PMFA		0.4	131
P(MFA-CO-MCN)	(20/80)	0.4	115
P(MFA-CO-MCN)	(57/43)	0.3	124
PMCN		0.3	120
P(MCN-CO-MCA)	(49/51)	0.6	120
P(MCN-CO-MCA)	(32/68)	1.2	130
PMCA		1.8	151,130
P(MCA-CO-MMA)	(46/54)	1.1	125
PMMA		0.9-1.2	105
P(MMA-CO-MFA)	(78/22)	0.7	106
PMFA		0.4	131

[a] See PE conditions of Table II.

The etch rate measurements for positive and negative-behaving
e-beam resists are found in Table V. It is apparent that the etch
resistance is lower the more sensitive the positive resist. The
exception would be PMCN, which exhibits better dry-etch resistance
than that which would be predicted based on e-beam sensitivity
alone. Where e-beam sensitivity and etch resistance are needed,
copolymerization becomes very important. This has been demonstrated for the MCN/MMA and MCA/MCN model copolymer systems in
references 9 and 10, respectively.

Dry-etch selectivities for several negative e-beam resists
are also listed in Table V. They are more resistant than the
positive e-beam resists of the Table except PMCN and the positive
photoresists, AZ2400 and PC 129. The positive-behaving vinyl
polymer resists tested are generally less resistant than the negative-behaving systems. This generality, however, does not hold
for the photoresists tested, as the data of Table VII verifies.

In general, the photoresists exhibit greater dry-process
resistance than the vinyl polymers of Table II. The greater dry-
etch resistances of photoresists is attributed to the aromatic
nature of the crosslinking agents, photoactive components, and
novolac resins (positive photoresists only). In addition, the

negative photoresist resins are known to be of the cyclized poly-(isoprene) type with varying degrees of unsaturation. It is this same compositional effect that leads to high thermal degradation resistance for the polymer resins. In the case of novolacs, thermal degradation does <u>not</u> yield monomer as for many of the vinyl resists. Thus, there is a good correlation between various types of data.

Table VII: Commercial photoresist etch rate ratios.

Resist	PE[a] Rate Ratio vs SiO$_2$	RIE[a] Rate Ratio vs SiO$_2$	Ion Milling[a] Rate Ratio vs SiO$_2$	Resist Tone
PMMA	0.9 – 1.2	1.2	1.2	+
Kodak 809	0.3	0.2	0.8	+
AZ2400/1350J	0.5	0.6	0.6	+
KTI II	0.4	0.5	0.7	+
PC 129	0.2	0.2	0.5	+
HPR 204	0.5	0.3	0.4	+
Merck Selectilux	0.3	0.3	1.5	−
Cop (Hunt)	0.6	0.6	0.8	−
Kodak 747	0.2	0.1	0.8	−
HNR 80	0.7	0.1	0.6	−

[a] See etching conditions of Table IV.

Summary

It is evident that dry-etch rates or their etch ratios can vary significantly for vinyl polymers with different side chain substituents. The aromatic vinyl polymer resists are the most resistant; this grouping also includes the novolac-based positive photoresists. Polymer resists with strongly bonded side chain groups like the α-fluorine or α-cyano acrylates, are also highly resistant. Halogenated poly(methacrylates), on the other hand, are significantly less resistant, except when the halogen is incorporated into the aromatic part of the polymer. Greater general etch resistance is observed for negative than for positive e-beam polymer resists. Greater dry-etch resistance is displayed by the photoresist systems.

Acknowledgment

The authors gratefully acknowledge technical discussions concerning this work with Dr. J. Lai.

Literature Cited

1. Thompson, L.F.; Kerwin, L.E., Ann. Rev. of Materials Sci.,
    1967, 6, 267.
2. Kirk, R.W. Chap. 9 in "Techniques and Applications of Plasma
    Chemistry"; Hollahan, J.R.; Bell, A.T. eds., Wiley, New York
    1978.
3. Taylor, G.N.; Wolf, T.M. Polym. Eng. and Sci., 1980,
    20, 1087.
4. Moreau, W.M. Mohawk Photopolymer Conference, June 1979.
5. Helbert, J.N.; Chen, C-Y.; Pittman, C.U.; Hagnauer, G.L.
    Macromolecules 1978, 11, 1104.
6. See references 20-23 of reference 5.
7. Pittman, C.U.; Chen, C-Y.; Ueda, M.; Helbert, J.N.;
    Kwiatkowski, J.H., J. of Polym. Sci: Chem Ed. 1980, 18, 3413.
8. Florin, R.F. in "Fluoropolymers", Wall, L.A. ed., Wiley,
    New York, Chap. 11 1972.
9. Stillwagon, L.E.; Doerries, E.M.; Thompson, L.F.; Bowden,
    M.J. Org. Coatings and Plastics Preprints 1977, 37, (2) 38.
10. Lai, J.H.; Helbert, J.N.; Cook, C.F.; Pittman, C.U. J. Vac.
    Sci. Technol. 1979, 16, 1992.

RECEIVED October 19, 1981.

Poly(*N*-alkyl-*o*-nitroamides)

A New Class of Thermally Stable, Photosensitive Polymers

S. A. MacDONALD and C. G. WILLSON

IBM Research Laboratory, San Jose, CA 95193

ABSTRACT: A new class of photosensitive, thermally stable polymers containing photo-labile aromatic amide linkages has been prepared. These polymers can be used to provide lithographic relief images for printing, etch masks for microcircuit fabrication and as contrast media for optical information storage.

At the present time, most of the positive photoresists used in the manufacture of microcircuits consist of a low molecular weight phenolic resin and a photoactive dissolution inhibitor. This composite system is not readily soluble in aqueous base but becomes so upon irradiation with ultraviolet light. When this resist is exposed, the dissolution inhibitor, a diazoketone, undergoes a Wolff rearrangement followed by reaction with ambient water to produce a substituted indene carboxylic acid. This photoinduced transformation of the photoactive compound from a hydrophobic molecule to a hydrophillic carboxylic acid allows the resin to be rapidly dissolved by the developer.(1,2,3)

While this two-component photoresist is very sensitive to light, it is also sensitive to heat, and this thermal lability imposes several limitations on the manufacturing process. The thermal instability of the composite resists can be attributed to two factors. First, the photoactive compound undergoes thermal decomposition and second, the glass transition temperature of the phenolic matrix is approximately 100°C. As a result of these two factors, all processing temperatures prior to the image exposure must be kept below 90°C to prevent significant decomposition of the diazoketone, and processing temperatures subsequent to image development must not exceed the flow temperature of the phenolic resin. While there are processing techniques to circumvent the latter difficulties, the manufacturing sequence would be greatly simplified with the development of a thermally stable and photosensitive resist.

In order to develop a material with these properties, the well-documented thermal characteristics of the aromatic polyamides were combined with the sensitivity of a photolabile protecting group. In 1973, Amit and Patchornik reported that N-substituted-ortho-nitroanilides are light-sensitive and undergo a photoinduced rearrangement to produce the corresponding carboxylic acid in excellent yield.(4) This general reaction is shown in Figure 1. The use of ortho-nitroanilides as a

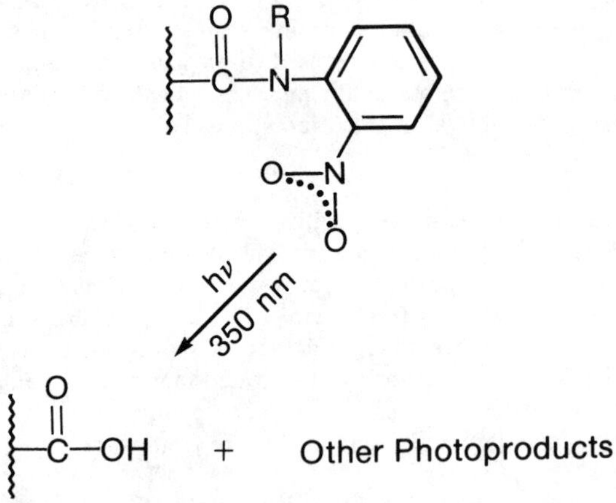

Figure 1.   Photochemistry described by Patchornik (4).

protecting group for carboxylic acids has not become a standard procedure, but the observation that a thermally stable N-arylamide bond can be converted into a photolabile linkage by the incorporation of an ortho-nitro functionality has led to the development of a thermally stable, photosensitive resist.

The materials that we have prepared contain a photosensitive anilide in the backbone of the polymer and are shown in Schemes I and II. When these polymers are exposed to ultraviolet light, they degrade in a manner that is analogous to the ortho-nitroanilides described by Patchornik. This resist system can be utilized in lithography because the photoinduced rearrangement not only reduces the molecular weight of the polymer but also converts a hydrophobic disubstituted amide into a carboxylic acid. Thus, after exposure to light, the irradiated areas can be dissolved in a basic developer solution, leaving the unexposed regions intact. This chemistry can be exploited to provide lithographic relief images for printing, etch masks for microcircuit fabrication and as contrast media for optical information storage.

Synthesis and Thermal Properties

We have prepared condensation polymers of both the A-A, B-B and A-B type which incorporate a nitroanilide into the mainchain of the polymer. The A-A, B-B monomers are 3,3'-dinitro-4,4'-di-N-methylaminodiphenyl ethers and commercially available aromatic diacylchlorides. The synthesis of the required diamine, which is produced in good yield, is outlined in Scheme III. The first two steps in the sequence have been reported by other workers.(5) This diamine has been condensed with both terephthaloyl and isophthaloyl chloride in dimethylacetamide to yield materials with a degree of polymerization from ten to twenty, as determined by vapor phase osmometry. Gel permeation chromatography in chloroform gave a number average molecular weight (relative to polystyrene) in the region of 2000 with a dispersivity of 2.4. The meta-linked polyamide (Scheme II) is readily soluble in most common organic solvents while the para-isomer requires solvents like dimethyl sulfoxide, dimethylformamide, or dimethylacetamide. Due to this difference in solubility, most of our lithographic studies were performed on the meta-isomer.

The thermal stability of these materials was examined by thermogravimetric analysis (TGA) and IR spectroscopy. As anticipated from their structures, the thermal properties of these polymers are far superior to those found in a typical diazoketone/phenolic resin resist. TGA (in air) of the material depicted in Scheme II shows that the polymer does not change in weight up to a temperature of 300°C. This resistance to thermal degradation was also verified by IR spectroscopy. The spectroscopic experiment was conducted by spin-coating the polymer onto a sodium chloride plate, recording the IR spectrum, heating the sample to 200°C in air for one hour and reexamining the spectrum. Within experimental error, the two spectra were identical.

The A-B monomer, 3-nitro-4-(N-methylamino)benzoyl chloride was prepared by the sequence outlined in Scheme I, which is similar to a procedure described in the literature.(6) This monomer undergoes spontaneous polymerization upon heating either in solution (dimethylacetamide) or in the bulk phase. When the melt polymerization is observed under a polarizing microscope, an unusual sequence of phase transitions occurs. First, the crystalline monomer is converted into a clear melt, gas evolution occurs, and then highly-birefringent needles grow out of the melt. The needles slowly lose order to ultimately provide a stable amorphous but birefringent solid. The resulting polymer has spectral and analytical characteristics consistent with

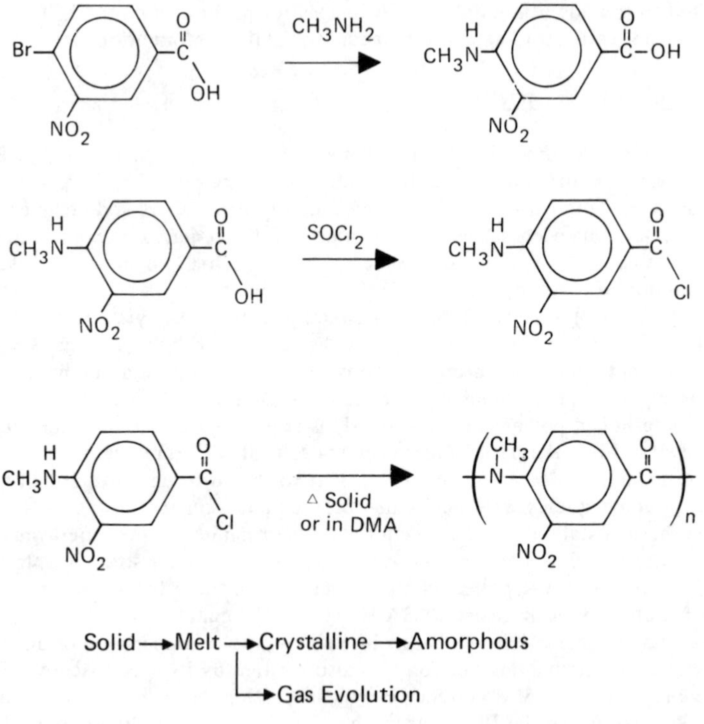

Scheme I. Synthesis and polymerization of 3-nitro-4-(*N*-methylamino)benzoyl chloride.

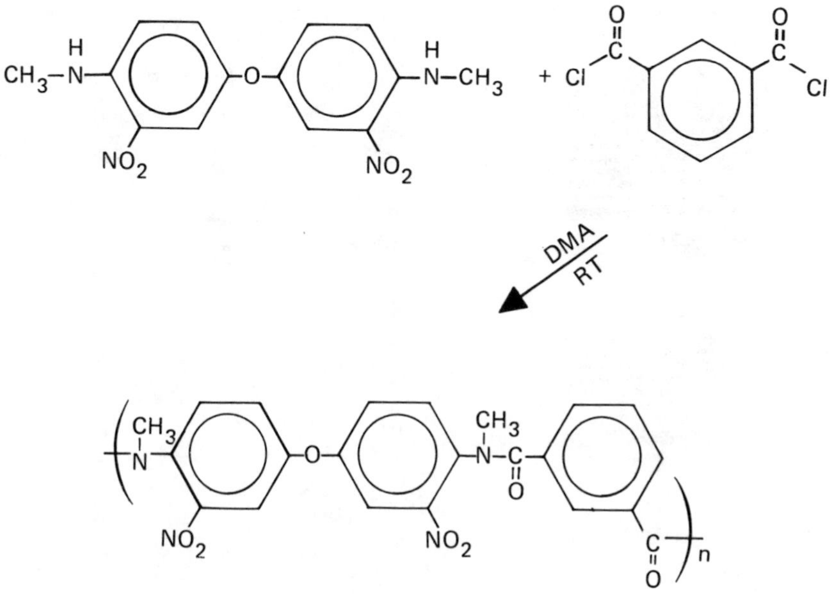

Scheme II.   Formation of *m*-polynitroanilide.

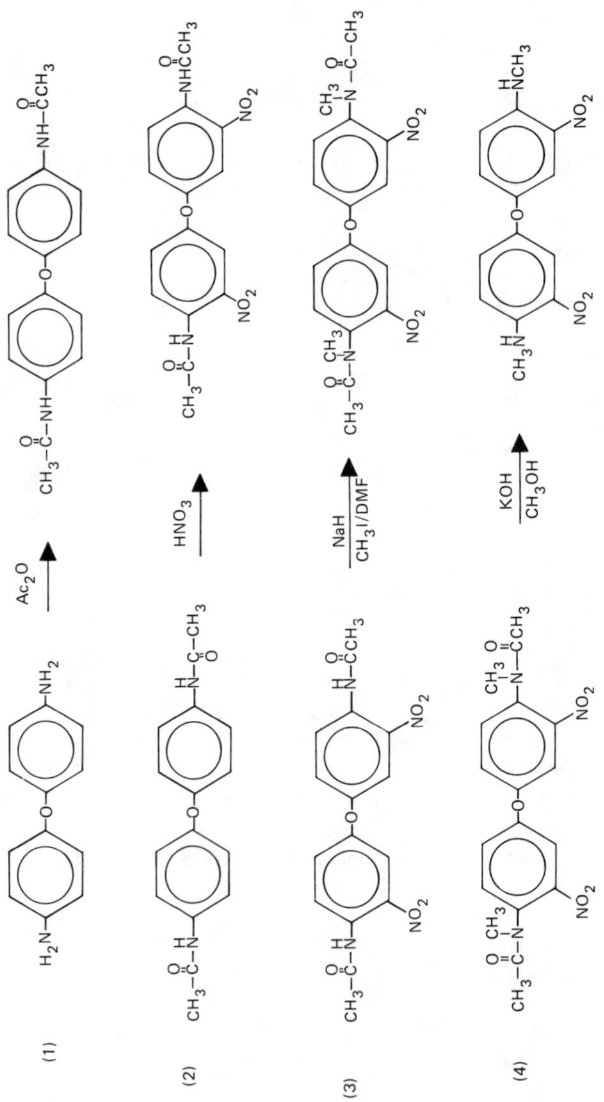

Scheme III. Synthesis of 3,3′-dinitro-4,4′-di-N-methylaminodiphenyl ether.

the A-B structure. The thermal properties of this polymer are very similar to those exhibited by the structure shown in Scheme II. Unfortunately, the A-B polymer is soluble only in hot dimethylacetamide or hot dimethylformamide, which severely restricts its processibility.

Application

The meta-linked polyamide (Scheme II) was dissolved in 1,1,2-trichloroethane at a concentration of 5 to 20% w/w and spin-coated on silicon wafers. The resulting films ranged in thickness from 0.06 to 3.5μ and after baking at 80°C were strongly adhering, uniform in thickness and continuous. The ultimate film thickness was controlled by varying the polymer concentration and spin-coating speed. These films were exposed on an Oriel illuminator containing a 1 Kwatt, Hg/Xe lamp with quartz optics and subsequently immersed in a tetramethylammonium hydroxide/dioxane developing solution. The exposed areas were dissolved to the substrate by the developer within 60 seconds without measurable loss of film thickness in the unexposed regions. Figure 2 shows high-contrast relief images that were produced in a 1μ thick film with an exposure of less than 100 mJ/cm^2.

Thin films (600Å) of these polymers can be imaged in a remarkable way by exposing them to UV light and then immersing them in acetone-water or methanol for a few seconds. This treatment renders the exposed areas transparent, providing contrast to the intense yellow color of the unexposed area. The image formed is not the result of a difference in film thickness since the exposed and the unexposed regions are identical in thickness after development. This process was used to produce high-resolution images by contact printing in a 600Å thick film on silicon to provide a contrast in reflected white light of approximately two.

At the present time we are investigating the quantum efficiency of the chemistry responsible for the bleaching phenomenon and continuing to study structural analogs of this system. We are also continuing to evaluate these polynitroanilides in several phases of microcircuit fabrication.

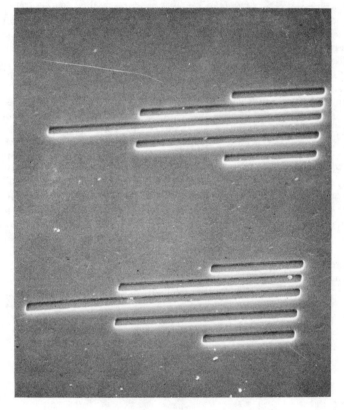

Figure 2.   A 5 × 10 μm and 5 × 5 μm line and space pattern in 1 μm of poly-nitroanilide resist.

Acknowledgments

The authors would like to thank T. Harvey and T. Schierling for technical assistance.

Literature Cited

1. DeForest, W. S., "Photoresist Materials and Processes," McGraw-Hill: New York, 1975, p. 19.
2. Bowden, M. J.; Thompson, L. F., Solid State Technol., 1979, 23, 72.
3. Deckert, C. A.; Ross, D. L., J. Electrochem. Soc., 1980, 127, 45C.
4. Amit, B.; Patchornik, A., Tetrahedron Letters, 1973, 2205.
5. Foster, R. T.; Marvel, C. S., J. Polymer Sci., 1965, Pt. A; 3, 417.
6. Foken, A. P.; Gerasenova, T. N.; Matoshena K. E.; Soklenko, V. E.; Ogneva, L. N.; Siberian Chemistry Journal, 1972, Ser. 4, 89.

RECEIVED October 19, 1981.

A New Approach to High-Resolution Lithography Based on Conducting Organic Charge Transfer Salts

E. M. ENGLER

IBM Research Laboratory, San Jose, CA 95193

Y. TOMKIEWICZ, J. D. KUPTSIS, R. G. SCHAD, V. V. PATEL, and M. HATZAKIS

IBM Watson Research Laboratory, Yorktown Heights, NY 10598

Conducting organic π-donor halide complexes such as tetrathiafulvalene bromide were discovered to act as electron beam resists which display a unique combination of useful properties. Exposure of sublimed films to an electron beam generates the neutral π-donor and the halogen which is subsequently lost from the film. Depending on exposure conditions, either negative (solvent developed) or positive (in-situ developed) resist images with a resolution of the order of 0.5μ can be generated. The strongly absorbing (uv-vis.) and highly conducting (~10/ohm-cm) films were found to become transmitting and insulating upon electron beam irradiation.

In the course of our research on organic metals, we discovered that certain of these materials can function as electron-beam resists for high resolution lithography with a combination of unique features that have no parallel among conventional resist materials.[1]

BACKGROUND

Since the discovery[2] in 1973 of metal-like conductivity in the charge transfer salt: tetrathiafulvalene-tetracyano-p-quinodimethane (TTF-TCNQ, 1-2), a host of new materials have been prepared displaying this interesting property. Widespread research on these materials has led to an improved understanding of the physics underlying the organic metallic state, and to a succession of molecular modifications which have enhanced these properties.[3]

0097-6156/82/0184-0083$05.00/0

1   2

For example, recently some selenium derivatives of TTF (i.e., TMTSF) were found to become superconductors around 1°K.[4]

Perhaps the two most fundamental requirements for high conductivity in organic charge transfer salts are:

- a segregated-stacked structure; that is, where the donor molecules stack one on top of the other in a uniform and closely spaced fashion, and the acceptor molecules do likewise. The conductivity in such a structure is highly anisotropic, with the metallic conductivity only in the stacking direction and poor conductivity in the other crystallographic directions.
- incomplete charge transfer; that is, where less than one electron is transferred per donor-acceptor pair in forming the salt. For example, in TTF-TCNQ the degree of charge transfer is ~0.6 electron per donor-acceptor pair. The donor stack can be considered as consisting of both neutral and radical cation states of TTF, while the TCNQ stack involves neutral and radical anion species.

As means of probing the role of charge transfer on transport properties, we investigated the solid state doping of neutral TTF with halogen acceptors.[5] Neutral TTF adopts a uniform stacked structure in the solid. Our intent was to deplete this filled-band insulator by solid state reaction with halogen as schematically illustrated in Figure 1. This would give rise to an incompletely charge transferred donor stack and consequently to high conductivity. By controlled reaction with halogen vapor, the conductivity of crystals of TTF could be systematically enhanced over 11 orders of magnitude.

The amount and uniformity of the solid state reaction of halogen with TTF was probed by the electron microprobe technique.[6] In this analytical method, low energy electron irradiation of a sample provides X-ray core level emissions, characteristic of the element and its relative concentration. Our initial analyses indicated a dramatic dependence of the halogen concentration with the energy of the electron beam. To probe this phenomenon further, TTF was reacted with bromine in solution to give a compound with the known composition:[7] TTF-$Br_{0.59}$. Electron microprobe studies of sublimed films of this material[8] indicated that at low electron irradiation energies, the correct bromine concentration could be determined. Figure 2a shows a BrL_α scan and the uniform distribution of bromine in the film. Exposure of a section of this

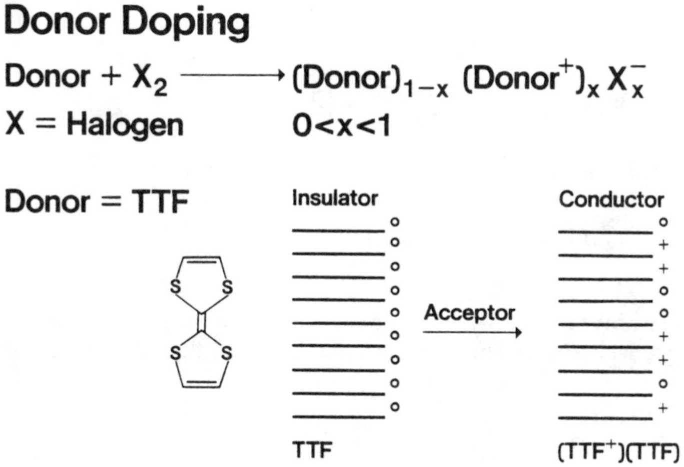

Donor Doping
Donor + X_2 → $(Donor)_{1-x}\ (Donor^{+})_x\ X_x^{-}$
X = Halogen
$0<x<1$
Donor = TTF
S
S
S
S
Insulator
Acceptor
Conductor
TTF
$(TTF^{+})(TTF)$

Figure 1. Halogen doping of neutral TTF.

a
b

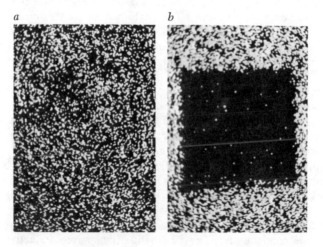

Figure 2. Electron microprobe BrL_α scan (1000×) taken on a $TTF\text{-}Br_{0.59}$ film: (a) before exposure and (b) after exposure of small area of film.

scan to a higher energy electron-beam was found to irreversibly deplete that region of bromine as shown when the film is re-examined at lower energy in Figure 2b. Electron microprobe analysis for sulfur showed no change in concentration between Figures 2a and 2b. These results suggested to us that the electron-beam was inducing a *reverse* electron transfer reaction as indicated in Eq. (1).

$$\text{TTF - Br}_{0.59} \quad \xrightarrow[\text{irradiation}]{\text{E} - \text{beam}} \quad \text{TTF} + \text{Br}_2 \uparrow \qquad (1)$$

$$|\ |\ |$$

$$(\text{TTF}^\circ)_{0.41}(\text{TTF}^+)_{0.59}(\text{Br}^-)_{0.59}$$

Loss of bromine produces dramatic changes in physical properties between exposed and unexposed areas and provides the basis for lithographic applications.

RESULTS

Typical lithography for the fabrication of microcircuits involves polymer resists where irradiation with light, E-beam or X-rays producing differential solubility changes between exposed and unexposed regions due to either chain cross-linking or scission.[9] The resist is usually applied to a substrate by spin-coating from solution. In the case of the conducting charge transfer salts we are dealing with molecular solids which can be deposited onto substrates by sublimation.

TTF halides are readily sublimable at temperatures ranging from 180-210°C and 0.5-5 Torr to give *smooth, adherent, glassy-like* films. This fortunate property was, in fact, essential for any possible use in lithography, and our investigations with other conducting donor halides indicated that these good film-forming properties are by no means general.

Scanning electron micrographs indicate that the grain size can be varied depending on the sublimation conditions. So far, we have been able to easily prepare films of TTF-Br$_{0.59}$ which show no discernable grain structure at 10000X.

The loss of halogen in the irradiated areas, as expected, leads to a dramatic drop in conductivity. The TTF halide films have conductivities on the order of 10-20/ohm-cm. Irradiation causes the conductivity to drop by over 9 orders of magnitude. High conductivity of the initial resist films is a useful property since it prevents static charge build upon E-beam irradiation which can lead to concomitant loss of image resolution.

The reverse electron transfer reaction (Eq. (1)) introduces significant change in the solubility characteristics between exposed and unexposed regions. TTF-Br$_{0.59}$ is hydrophilic and preferentially dissolves in polar solvents such as alcohols, while neutral TTF is hydrophobic and dissolves in nonpolar solvents such as hydro- or halo-carbons. Therefore, in principle, either positive or negative resist images should be achievable depending on the choice of developer solvent. In practice, however, we have found that these π-donors tend to easily cross-link under electron-beam irradiation, so that only negative resist images are obtained by solvent development. Figure 3 shows typical negative resist images for a TTF-Br$_{0.59}$ film which was exposed to an electron-beam dose of 10^{-5} coulomb/cm^2 and developed by washing with methanol (R.T., 30 sec). Lift-off of the negative images is accomplished by treatment with a dilute base such as aqueous hydrazine.

While positive resist images do not appear readily achievable by solvent development, when the electron dose is increased to 10^{-4} coulomb/cm^2, the neutral TTF that is formed in the exposed areas sublimes due to local heating before significant cross-linking can occur. The resulting positive image is, therefore, generated in-situ. Figure 4 shows a pattern on TTF-Br$_{0.59}$ resist developed in this manner. In this mode, the electron beam functions not only as a cause for reverse electron transfer, but also as a heat source. Since neutral TTF has a low sublimation temperature, sensitivity of at least 10^{-4} coulomb/cm^2 can be achieved with a beam current of 2×10^{-6} amp.

π-Donor cross-linking and sublimation are competing processes, and current density (not charge density) determines which of them dominates. While cross-linking can occur in the presence of the halogen, the essential requirement for the sublimation is the presence of the neutral donor, namely, elimination of the halogen. In Figure 5, the relative change in the halogen concentration is plotted as a function of exposure time for two different current densities. It is clear that while a beam of diameter 4μ causes a complete exclusion of bromine within 25 sec, a beam of 50μ diameter, cannot reduce the bromine concentration below 58% of its original value. Moreover, when the areas exposed to the low current density are re-exposed to the high current density, no further reduction of bromine concentration could be achieved--the most probable reason being that the bromine is locked into the cross-linked TTF network. Thus, the sensitivity of the resist, at least in its positive mode, is current density dependent. It should be noted that an increase of the beam current from 2×10^{-7} to 2×10^{-6} amps increases the sensitivity by an order of magnitude.

The electron-beam induced reverse electron transfer reaction appears to be a rather general phenomenon for the class of conducting π-donor halide complexes. For example, tetrathiatetracene (TTT,3) and tetraselenafulvalene (TSeF,4, i.e., the Se analog of 1) halides also undergo loss of halogen on irradiation. Furthermore, the halogen can be either Cl, Br or I. Some changes in sensitivity have been observed for TTF halides, with the TTF-chloride being about 20% more sensitive than the bromide.

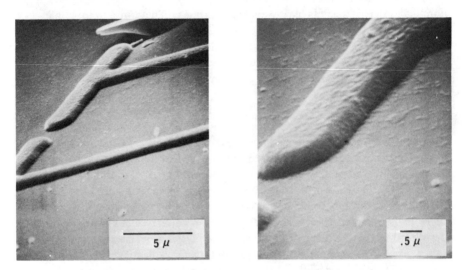

Figure 3.   SEMs of negative resist images of TTF-Br$_{0.59}$.

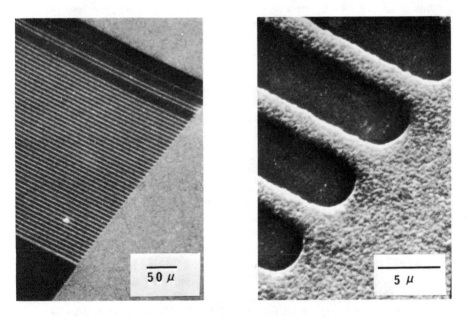

Figure 4.   SEMs of positive resist images of TTF-Br$_{0.59}$.

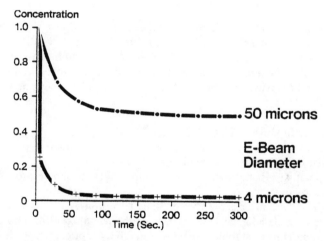

Figure 5. The relative change in the halogen concentration as a function of exposure time for two current densities differing by 2.5 orders of magnitude.

3   4

CONCLUSION

At this stage, the potential of E-beam induced reverse electron transfer in conducting organic charge transfer salts for lithographic applications is unclear. However, these materials do possess a rather unique combination of properties that may be of considerable value in future applications where traditional resist materials may be unsuitable. Some of the key features of these new resists are summarized below:

- conducting resist films;
- negative and positive images possible from same material;
- resist deposited by sublimation;
- in positive resist mode, completely dry and very simple process (in-situ image development);
- high resolution possible (better than 1 micron);
- ease of molecular tailoring of resist (donor, halogen) to vary properties (e.g., high temperature stability, E-beam sensitivity, optical properties, conductivity, etc.).

Two approaches appear promising for enhancing the sensitivity of these materials to E-beam irradiation. The first involves the use of lightly halogen-doped donors (as discussed earlier for TTF, Figure 1) so that the amount of halogen to be removed is significantly less. The other approach is based on our studies which indicate a dependence of in-situ film development (positive mode) on how the E-beam dose is applied. E-beam heating of the substrate is apparently important in facilitating donor sublimation. We are currently evaluating the effect on sensitivity when the temperature of the substrate is varied.

Literature Cited

1. Aspects of this work were previously published: Y. Tomkiewicz, E. M. Engler, J. D. Kuptsis and R. G. Schad in The Physics and Chemistry of Low Dimensional Solids, ed. L. Alcacer (D. Reidel Publishing Co., Boston, 1980), p. 413.
2. J. Ferraris, D. O. Cowan, V. Walatka, Jr. and J. H. Perlstein, J. Amer. Chem. Soc. 95, 943 (1973); L. B. Coleman, M. J. Cohen, D. J. Sandman, F. G. Yamagishi, A. F. Garito and A. J. Heeger, Solid State Comm. 12, 1125 (1973).
3. For a recent survey of this field see: Molecular Metals, ed. W. E. Hatfield (Nato Conference Series, Plenum Press, New York, 1980).
4. D. Jerome, A. Magaud, M. Ribault and K. Bechgaard, J. Phys. Lett. 45, L-95 (1980).
5. Y. Tomkiewicz, E. M. Engler, B. A. Scott, S. J. LaPlaca and H. Brom, p. 43 in reference 3.
6. For example, this technique was successfully employed in determining the composition of organic alloys: E. M. Engler, B. A. Scott, S. Etemad, T. Penney and V. V. Patel, J. Amer. Chem. Soc. 99, 5909 (1977).
7. For details on the solid state properties of this material see: B. A. Scott, S. J. LaPlaca, J. B. Torrance, B. D. Silverman and B. Welber, J. Amer. Chem. Soc. 99, 6631 (1977).
8. Elemental and X-ray analyses of these films provided an identical composition as the unsublimed TTF-$Br_{0.59}$.
9. For a recent review see: M. J. Bowden, CRC Critical Reviews in Solid State and Materials Science 8, 223 (1979).

RECEIVED November 12, 1981.

Polyimide for Multilevel Very Large-Scale Integration (VLSI)

GAY SAMUELSON

Process Technology Laboratory, SRDL, Motorola, SG, Phoenix, AZ 85008

Multilevel structures consisting of alternating metal and dielectric layers are necessary to achieve interconnection in high
density or VLSI circuits using either MOS or bipolar technology.
The function of the interlevel dielectric of the multilevel structure is three-fold: (1) it must provide planarization of underlying topography while allowing high resolution patterning of via
holes necessary for contact between metal layers, (2) it must provide insulation integrity, and (3) it must contribute minimally to
device capacitance.

A likely candidate for the role of interlevel dielectric is
polyimide on the basis of its relative purity and planarizing spin-
on application. In fact, planarizing metal with polymer or so-
called PMP technology was pioneered by Hitachi to develop two
metal level transistors (1,2,3). More recently, several other
companies, TI (4), and IBM (5) have reported use of polyimide for
multilevel interconnect systems. T. Herndon and R. Burke have
reported a process for constructing polyimide-aluminum multilevel
64K MNOS memories (6).

The present work is a report of the properties of polyimide
which define functionality as an interlevel dielectric/passivant.
Thus, the planarizing and patterning characteristics and electrical characteristics of current vs voltage, dissipation, breakdown field strength, dielectric constant, charge and crossover
isolation are discussed in addition to the reliability-related
passivation properties.

Experimental

Baseline Process. DuPont PI2545, PI2555 and Hitachi PIQ as
received from the manufacturer, were spun in a class 100 clean
room environment at appropriate spin speeds to achieve 0.5 - 6 μ
film thickness. The silicon wafer substrates were pre-spun
(5K rpm, 30") with 0.05% DuPont VM651 (γ-amino propyltriethoxy
silane) adhesion promoter in 95/5 (v/v) methanol/H_2O. The polyimide film cast on the silane-coated silicon wafer was pre-baked

30 min at 130°C after which positive resist such as KTI-809 was
spun, soft-baked and exposed. Resist development and concomitant
polyimide etch occurred on spray application of a positive resist
developer like DE-3. Acetone was used as a resist strip and methanol was used as a final rinse. The patterned polyimide film was
cured as follows:

> 1 hr   200°C
> 1 hr   300°C
> 15 min 400-450°C

Dry Etch Conditions. Fully cured polyimide films were used
for all dry etching. The degree of desired resolution determined
the mask. For larger geometries (e.g., 4 - 5 μ), hard-baked KTI-
II resist was chosen. Since the etch rate of resist was twice
that of fully cured polyimide, the resolution achieved with this
system was limited by the thick (2.7 μ) resist necessary to maintain mask integrity for a 1.2 μ thick polyimide film. For smaller
geometries (e.g., 0.5 - 3 μ), a modification of the Bell Labs
technique was used (7). Plasma enhanced (PE) SiO_2 or SiN was
deposited at a thickness of 1200 Å on 1.2 μ of fully cured polyimide. The inorganic film was plasma etched and subsequently
used as a mask for reactive ion etching (R.I.E.) or reactive ion
milling (R.I.M.) of polyimide.

Three different dry etch techniques were investigated: isotropic O_2 plasma etching in a Tegal 200 reactor, R.I.E. in a par-
allel-plate in-house modified Tegal 400 reactor and R.I.M. in a
Veeco, Model RG-830. The conditions of operation for each system
were as follows where time is the time to etch 1.2 μ of fully
cured polyimide.

> i) Tegal 200 - 1.2 torr O_2 press., 300 watts, 5 min
> ii) Modified Tegal 400 - 100 μ O_2 press., 350 watts, 9 min
> iii) Veeco, RG-830 - 9 x 10^{-4} torr O_2 press., 15° angle
>   of incidence of the incoming beam, 500 V accelerating
>   voltage, 0.55 ma/cm^2 current density, 20 min

Electrical Measurements. Dielectric and I-V characteristics
were determined on simple guard ring-dot MIS structures consisting
of Al - polyimide - degenerate silicon of resistivity 0 - .02 Ωcm.

The test structure for C-V characteristic determination was
similiar except that resistivity of the silicon wafer substrate
was 6 - 12 Ωcm.

I-V characteristics were determined using a Keithley 616
electrometer, a Kepco model BPO 500M bipolar high voltage power
supply and a Fluke 8502A high resolution DVM. C-V characteristics
and dielectric properties were determined using an HP 4275A LCR
meter.

The pinhole density of polyimide was assessed by a statistical evaluation of shorts using an TiWAu - polyimide - TiWAu multilevel structure where each die contained 3275 crossovers of first
and second metal. The probability of good crossovers was taken as

$(1-P)^N$ where P is the probability of one bad crossover and N is
the total number of crossovers. The probability of good crossovers was determined experimentally from the total number of open
die $\div$ the total number of die probed.

Results and Discussion

The properties of polyimide which pertain to the fulfillment
of each functional requirement of an interlevel dielectric/passi-
vant will be discussed in turn.

Planarization and Patternability. Polyimide, because of its
spin-on application, is an ideal choice for planarizing underlying
topography. An example is provided in Figure 1 which shows a
scanning electron micrograph (SEM) of a cross section through the
bird's beak created at the periphery of an oxide isolation area
of a bipolar device. The planarizing effect of 1.6 μ PI2545 on
the 6000 Å step created by the beak, is evident. According to
Rothman (8), because of geometry effects, it is probably impossi-
ble to totally planarize but step coverage is vastly improved with
an underlying coat of polyimide.
Patterning the planar polyimide is highly process dependent
in terms of the resolution and wall slope achieved. Wet chemical
etch and barrel O_2 plasma etch are isotropic processes producing
sloping via walls and a resolution limit of 3 - 5 μ. An example
of isotropic O_2 plasma etching is provided in Figure 2 where a
5 x 5 μ via is shown etched in 1.5 μ PI2545 overlying large grain
size aluminum. Using a SiN or SiO_2 mask, directional etch techniques such as R.I.E. or R.I.M. can provide resolution to $\leq$ 1 μ.
Via walls, in this case, are essentially vertical which may require innovations in subsequent metallization to avoid step
coverage difficulty. This particular problem is unique to VLSI
where high resolution patterning requirements in dielectrics,
organic or inorganic, dictate use of directional etch techniques.

Insulation Integrity. Insulation integrity is a function of
an interlayer dielectric/passivant defined by specific electrical,
mechanical and passivation properties. The D.C. electrical property of interest is the I-V characteristic which is used to deduce
conductivity and breakdown field strength. The corresponding A.C.
electrical property is dissipation factor. The pertinent mechanical and passivation properties are, respectively, pinhole density
and performance rating as a diffusion barrier to Na^+ and H_2O.
Both bulk and surface I-V characteristics were determined for
PI2545, PI2555 and PIQ. A representative bulk current density J,
vs electric field E or $\sqrt{E}$ is shown in Figure 3 for 0.5 μ thick
PIQ. Log J appears to be a non-linear function of either E or $\sqrt{E}$
in the broad field range investigated although in the narrow range
of 5 x 10^4- 5 x 10^5 V/cm, log J vs. $\sqrt{E}$ is apparently linear as has
been reported (9). The overlapping multiple traces representing

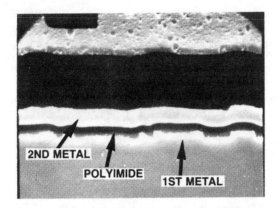

Figure 1.   SEM of the cross section through a 6000 Å bird's beak showing the planarizing effect of 1.6 μm PI2545.

Figure 2.   SEM of a 5 × 5 μm via in 1.5 μ PI2545.   Isotropic O₂ plasma etch conditions were used.

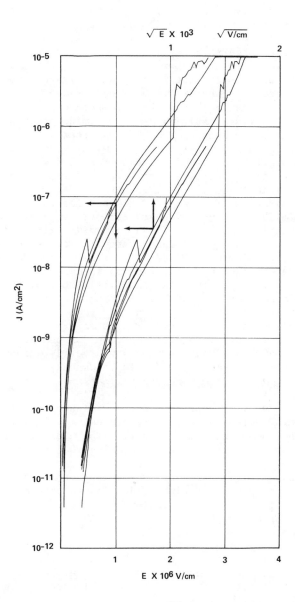

Figure 3.   Bulk J vs. E or $\sqrt{E}$ for 0.5-μm thick PIQ.

different wafer areas indicate good electrical uniformity for this
thin film. However, below 1 μ, electrical uniformity was shown to
be polyimide chemistry and/or solvent dependent. At a typical use
field condition of 5 x 10^5 V/cm, conductivity in polyimide is
closely similar to that for thermal SiO_2 (i.e., $\sim 10^{-16}$ $\Omega^{-1}cm^{-1}$)
but shifts several orders of magnitude larger than thermal SiO_2
(to ~ 3 x 10^{-13} $\Omega^{-1}cm^{-1}$) at higher fields such as E = 2 x 10^6 V/cm.
Surface I-V characteristics are indicated in Figure 4 for a poly-
imide surface before and after Ar^+ backsputter used as a pre-metal
clean. Ar^+ backsputter has been shown to remove 70 - 140 Å
organic residue in vias resulting from polyimide processing (6).
From Figure 4 it is evident that Ar^+ backsputter activates the
polyimide surface such that it becomes ohmically conductive with
a measured sheet resistivity of approximately 1.5 x 10^{13} $\Omega/\square$.
Furthermore, the electrically active surface does not readily
decay to its original background I-V characteristic as indicated
in the I-V scan Figure 4(c), made 72 hours following Ar^+ backsputter. A similar surface I-V characteristic and calculated
sheet resistivity were observed for a polyimide film following
wet chemical etch if there was inadequate removal of resist and/or
developer. The problem was solved by increasing the resist strip
time in acetone and subsequently, rinsing in methanol.

Another property pertinent to the fulfillment of the requirement for insulation integrity is dissipation factor (D). D is a
sensitive indicator of cure conditions (10) and its value is,
consequently, dependent on the cure regime as illustrated in
Figure 5 where D (1 MHz) is plotted vs time for temperatures in
the range 200 - 400°C. Such a series of curves is unique for a
particular polyimide chemistry and film thickness range. D decreases with time at T <300°C due to solvent and H_2O release and/
or imidization. To achieve the absolute minimum value of D however, it is necessary to bake at 300<T<450°, the specific temperature range being polymer chemistry dependent. This latter bake
phase drives out H-bonded water (11,12). The minimum D (1 MHz)
achievable under these conditions is 0.003 - .007 for all three
polyimides investigated.

The breakdown field strength is a thickness dependent property, a probable reflection of pinhole density variation with thickness. For a typical value of interlevel dielectric film thickness
(1 - 2 μ), the breakdown field strength is 1 - 2.5 x 10^6 V/cm
which is adequate for most applications.

Pinhole density is another property of interest in defining
insulation integrity. It was indirectly assessed from the number
of shorts in a statistical number of probed die where the die was
a multilevel test structure consisting of TiWAu-polyimide-TiWAu
with 3275 crossovers of first and second metal per die. The results indicated that the probability of a short in a crossover for
1.2 μ thick PI2545 was 1 in 133,333.

Passivation quality is yet another aspect of insulation integrity. The reliability of phosphosilicate glass (PSG) and poly-

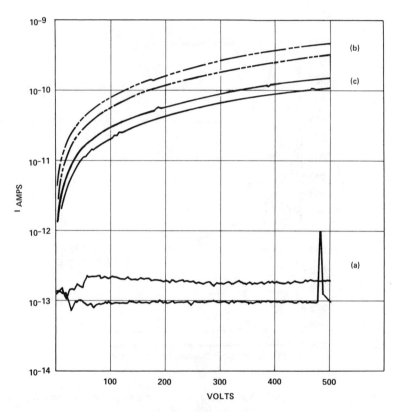

Figure 4.   Surface I–V characteristic of a polyimide film. Key: a, before Ar⁺ back-sputter; b, immediately after Ar⁺ backsputter; and c, 72 h after Ar⁺ backsputter.

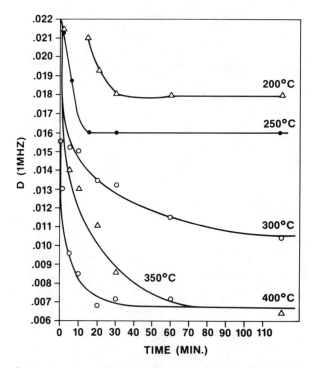

Figure 5.   Dissipation factor D *at 1 MHz vs. time at a specified temperature for a 1-μm thick polyimide film.*

imide passivated linear devices was determined from I-V characteristics of statistically significant numbers of devices following
severe PTHB test (i.e., 15 psi, 120°C, 100% relative humidity, and
30 V bias). Two coat (3 μ) polyimide passivation provided almost
twice the mean time to failure of 1 μ thick PSG passivation.
Polyimide protection against high humidity (13,14) and Na+ diffusion (15) has been reported previously.

   Low Interlayer Capacitance. A multilevel structure consisting of a metal-dielectric-metal sandwich is a capacitor whose
capacitance is determined by the dielectric constant ε and total
charge of the dielectric.

   The dielectric constant like the dissipation factor D is dependent on the cure conditions although, less sensitively. The
optimum cure conditions defined as those producing the minimum
possible D were determined for a 1 μ thick polyimide film by measuring D vs time at a specific temperature in the range 200°C -
450°C. D and ε were then measured on variously thick films all
cured under the same conditions considered optimal for a 1 μ thick
film. Figure 6 shows that the 1 μ cure conditions are adequate
for the thickness range investigated, namely 0.4 - 2.5 μ, since
little variation in either D or ε is observed over that thickness
range. Thus, the cure conditions are now completely defined for
this particular polyimide in the thickness range of interest and
the dielectric constant (1 MHz) can be confidently stated as being
3.2 - 3.5. The same ε was measured for all three polyimides
under conditions of optimum cure.

   C-V analysis of PI2545 and PIQ revealed varying degrees of
hysteresis and large flat band voltage shifts of varying sign
depending on the polyimide measured. A typical C-V trace for
PI2545 is shown in Figure 7. The flat band voltage is shifted
extremely in the direction of negative voltage indicating positive
charge in the film and the sense of hysteresis suggests injection
at the semiconductor/polyimide interface. The C-V profile for PIQ
on the other hand, indicated less hysteresis but extreme flat band
voltage shifts in the positive field direction. For calculating
charge, it is assumed that the hysteresis is due solely to injected charge and it as well as the intrinsic space charge of the
polyimide resides at the polyimide semiconductor interface at the
maximum excursion flat band voltage, $V_{FB}(2)$ in Figure 9. Thus
$V_{FB}(2)$ is used to calculate total charge, $V_{FB}(1) - V_{FB}(2)$ is used
to calculate injected charge and the difference is intrinsic or
space charge. The results of such calculations are indicated in
the following table.

Table I
Total, intrinsic and injected charge in
PIQ and PI2545

	PIQ	PI2545
Total Charge	-2.54×10^{11} cm^{-2}	$+8.30 \times 10^{11}$ cm^{-2}
Intrinsic Charge	-2.54×10^{11} cm^{-2}	$+6.0 \times 10^{11}$ cm^{-2}
Injected Charge	---	$+2.23 \times 10^{11}$ cm^{-2}

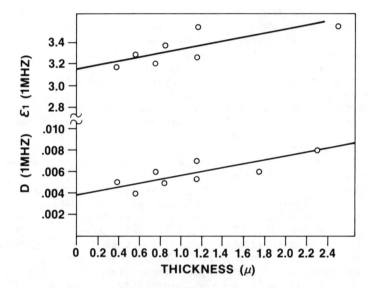

*Figure 6.   Real part of the dielectric constant ϵ_1 (1 MHz) and dissipation factor D
(1 MHz) as a function of polyimide thickness.*

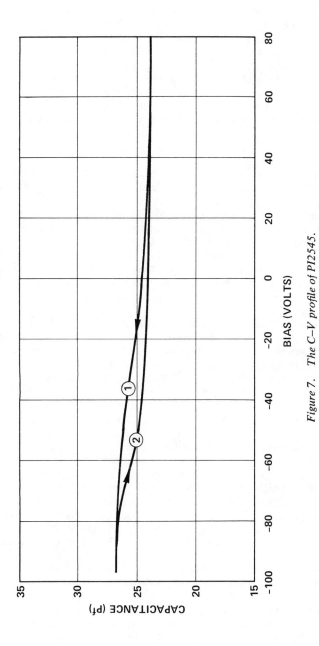

Figure 7. The C–V profile of PI2545.

The quantity of charge is considerable and could pose a reliability problem for polyimide/oxide underlying second level leads
where charge induced inversion of silicon is a potential hazard.
Certainly, large threshold shifts have been observed in Al-poly-
imide-phosphorus doped SiO_2-silicon transistors ($\underline{16}$). Several
approaches can be taken to prevent polyimide charge induced inversion of underlying silicon. One approach consists of increasing the thickness and decreasing the effective dielectric constant
of insulators intervening between polyimide and silicon ($\underline{17}$). A
second approach would require the use of field implants or guard
rings whose doping levels would essentially preclude inversion.

Summary

The properties of polyimide which particularly address the
functional requirements of a VLSI interlevel dielectric/passivant
are numerous.
Planarization using polyimide while probably never complete
because of topography dependence is, nevertheless, much better
than that achieved with conformal inorganic dielectrics. Patterning capability is excellent. Larger dimensions (3-5 μ) are patterned with good definition by isotropic wet or dry etch processes
using photoresist as a mask. Smaller dimensions (<3 μ) are best
achieved using a directional etch technique with a hard PE SiN or
Pe SiO_2 mask.
The electrical properties of the polyimides investigated are
all consistent with good interlevel dielectric performance. The
room temperature I-V characteristic indicates non-linearity of
current density with field or square root of field. At typical
field use conditions of 5×10^5 V/cm, polyimide conductivity is
$\sim 10^{-16} \Omega^{-1} cm^{-1}$ which is similar to that of thermal SiO_2. At higher
fields such as 2×10^6 V/cm, conductivity of polyimide increases
several orders of magnitude over thermal SiO_2 to $\sim 3 \times 10^{-13}$ Ω^{-1}
cm^{-1}.
Ar^+ backsputter as a standard pre-metal clean or incomplete
resist and/or developer removal in wet patterning results in an
ohmically conducting surface of sheet resistivity 1.5×10^{-13} $\Omega/\square$.
Otherwise polyimide film surfaces are not measurably conductive.
The dielectric properties of polyimides are highly dependent
on cure conditions. Having achieved optimum cure however, the
measured dissipation factor at 1 MHz is 0.003 - 0.007 and the
measured dielectric constant a 1 MHz is 3.5 for the three polyimides investigated.
The breakdown field strength at a standard interlevel application thickness of 1 μ, has an acceptable value of $1 - 2.5 \times 10^6$
V/cm.
Pinhole density was indirectly assessed by measuring shorts
in a series of crossovers of first and second metal with intervening polyimide. For 1.2 μ thick PI2545, the probability of a
short was estimated at 1 in 133,333 crossovers.

C-V analysis of MIS test structures using polyimide as the
insulator, indicates charge injection from the silicon and relatively large quantities of space charge or intrinsic charge whose
sign is dependent on polymer chemistry. The charge may limit applications of polyimide to interlevel functions where charge is
less critical and prohibit its use in applications with greater
semiconductor proximity.

Finally, the passivation properties of polyimide are superior
to phosphosilicate glass under conditions of severe PTHB testing.

In conclusion, DuPont PI2545 and PI2555 and Hitachi PIQ are
all good candidates for interlevel dielectric/passivant application in multilevel VLSI.

Acknowledgment

I would like to express thanks to Kathleen Ginn for excellent
technical assistance in preparing patterned polyimide films.
Thanks are also extended to Dr. Glenn Shirley and Dan McGuire for
making electrical measurements and especially to Dr. Glenn Shirley
for helpful theoretical discussion. Finally, I would like to
thank Melinda Gibbs for typing the manuscript.

Literature Cited

1. Saiki, A.; Mori, T.; Ome; Harada, S.; Hachioji; Sato, K.
    U.S. Patent 3,846,166, Nov. 5, 1974.
2. Mukai, K.; Saiki, A.; Yamanaka, K.; Harada, S.; Shoji, S.
    IEEE Journal of Solid State Circuits, 1978, SC-13 (4), 462.
3. Saiki, A.; Harada, S.; Okubo, T.; Mukai, K.; Kimura, T.
    J. Electrochem. Soc., 1977, 124, (10), 1619.
4. Shah, P.; Laks, D.; Wilson, A. IEDM, Dec. 1979, 204.
5. Picciano. Electronic News, Oct. 1, 1978.
6. Herndon, T.; Burke, R.L. Kodak Microelectronics Seminar,
    New Orleans, Oct. 1979.
7. Moran, J.M.; Maydan, D. The Bell System Technical Journal,
    1979, 58, 1027.
8. Rothman, L.B. J. Electrochem. Soc., 1980, 127, (10), 2216.
9. Zielinski, L.B. ESC Meeting, Seattle, 1978. Abstract No.
    116,274.
10. Gregoritsch, A.J. 14th Annual Proceedings of Reliability
    Physics Symposium, 1976, 228.
11. Samuelson, G. unpublished results.
12. Wilson, A.; Laks, D.; Davis, S.M. ACS Organic Coatings and
    Plastic Chemistry, 1980, 43, 470.
13. Miller, S. Circuits Manufacturing, April 1977, 39.
14. Saiki, A.; Mukai, K.; Takahashi, S.; Yamanaka, K.; Harada, S.
    "Reliability of Semiconductor Devices Using a Resin Insulat-
    ion Structure," Central Research Laboratory Technical
    Bulletin, Hitachi, Ltd.

15. Harada, S.; Sato, K.; Saiki, A.; Kimura, T.; Okubo, T.;
     Mukai, K. J. of Japan Soc. Appl. Phys., 1975, 44, 297.
16. Brown, G. 19th Annual Reliability Physics Symposium Proceed-
     ings, 1981, in press.
     See also this volume.
17. Snow, E.H.; Dumesnil, M.E. J. of Applied Phys., 1966, 37,
     2123.

RECEIVED October 19, 1981.

Polyimide Coatings for Microelectronic Applications

Y. K. LEE and J. D. CRAIG

E. I. du Pont de Nemours, Inc., Marshall Research and Development Laboratory,
Philadelphia, PA 19146

The use of polyimide(PI)coatings as dielectrics and/or for
passivation for semiconductors and thin film hybrids has become
increasingly important.the principle reasons are: 1)Polyamic acids,
the precursors of polyimides, are solvent soluble to give viscous liquids that can be spun onto a wafer to create a relatively
planar surface that is suitable for the next level metallization.
Multilevel construction is essential to the development of very
large scale integration. (VLSI) 2)The cured polyimide coatings
are tough and resilient. They give excellent mechanical protection. 3)Polyamic acid coating solutions can be spun, exposed,
and etched with existing equipment.

Vapor deposited silicon nitride(Si_3N_4)and silicon dioxide
(SiO_2) are currently used in almost all microelectronic devices
as the dielectric or protective layer. They have the advantages
of purity, chemical inertness and low permeability to water
vapor. It is difficult, however, to apply a glass coating without
some mechanical defects such as pin holes, microcracks, etc.
Polyimides are more permeable but can be applied in much thicker
coats without the cracking problems. Thus, depending on the application, PI coatings can be used as a replacement or supplemen-
tary coating to the currently used SiO_2 and Si_3N_4. Currently,
high purity polyimide coatings are being used or are under development in the fabrication of semi-conductor devices in the
following areas: a) as a protective overcoat b) as an interlayer dielectric for multilevel devices c) as an alpha particle
barrier d) as an ion-implant mask.
a) Protective Overcoat - the presence of pinhole defects in a
passivation layer for an integrated circuit may be reduced or
eliminated by the use of a polyimide topcoat over the commonly
used passivation coatings of phosphosilicate glass (PSG) or silicon nitride. The chance of two defects from two separate coats
occurring at the same position is highly unlikely. Therefore,
the use of polyimide as a second protective coating leads to
improved yield and enhanced reliability.
b) Interlayer Dielectric - deposited SiO_2 can be used as an interlevel dielectric in multilevel structures on monolithic integrated circuits (1)(2)(3). However, with this construction,

0097-6156/82/0184-0107$05.00/0

reliability and yield become problems due to the topography of metal edges and via holes or oxide windows for interlayer connections.
Polyamic acid is a viscous liquid which will flow into the cavities and produce a relatively flat surface for the next level metallization. Since multilevel metallization may be the key to the
to the construction of very large scale integration(VLSI) devices,
this application probably provides the greatest value-in-use
among all polyimide applications.
c) Alpha Particle Barrier-Alpha radiation emitted by trace amounts
of naturally occurring thorium and uranium isotopes in packaging
materials is a source of nondestructive soft error problems in charge coupled devices and dynamic memories(4). As the devices geometries shrink and critical charge levels diminish, this problem
becomes much more acute. Thus, the chance for a soft error to occur
increases greatly from 16K to 64K RAM.
       A coating of polyimide with a thickness of no less than 3
mil over the memory chip will practically eliminate this problem.
PI coatings can be applied by dispensing via the syringe technique
or by using a cured film with a PI adhesive. d) Ion-Implant Mask-
The current approach in making ion-implant masks involves the use
of photoresist or metal masks. However, the use of conventional
photoresist restricts the energy and ion beam density to low levels
because high temperature will cause the resist to deform and flow.
The use of metal masks is a complex and expensive process involving the use of expensive equipment. The heat resistance and etcha-
bility of polyimide either by dry or wet process makes it feasible
for this application. G.Samuelson(5) and T.Herndon(6) reported
that with reactive ion etch the polyimide film gave vias with
verticle wall, a highly desirable characteristic for ion implant-
ation.
       In this paper, we shall discuss the processing steps used
in a typical PI coated wafer. The chemistry that is relevant to
processing and film properties will also be discussed with special
emphasis on adhesion, cure cycle, and thermostability.
EXPERIMENTAL
       The rate of imidization or degree of cure curves were dev-
eloped using a Nicolet 7000 Fourier-Transform Infrared instru-
ment. This FTIR has subtraction and time-lapse scanning capabil-
ities. Point by point measurements were taken in cure studies run
on KBr plates optimized for film thickness and reproducibility.
Curves generated by a dynamic method i.e.: no cure to full cure
in-situ, by increasing the temperature in steps, compared favorably
with the more conventional oven bake time and temperature method.
Substraction spectra were also generated showing the solvent loss
and conversion to imide. Films cast on silicon wafers have
been examined in the same manner with no difficulty using FTIR.
       The thermal gravimatic analysis (TGA) data was generated
using a Du Pont 1090 Thermal Analyzer. Both free films which had
been stripped off wafers and the film-coated wafers were examined
in air and nitrogen. The film thickness was twelve (12)microns and

the films had been baked 60 min.x300°C in air prior to examination in the Du Pont 1090 Thermal analyzer.
     The study of retained NMP in the films was conducted using a
Du Pont DP-102 Mass Spectrometer. These films were heated in a
tube-type pyrolysis furnace and the amount of NMP given off detected and quantitized against a calibration curve. Pyrolysis
conditions were 750°Cx2 seconds with quantitation done on mass
99 molecular ion. A CDS-190 Pyroprobe was used for the pyrolysis
study.
     The following experimental procedure was used to study adhes-
ion. Clean, virgin wafers with surfaces of polished silicon,
silicon oxide, silicon nitride, and phosphorous doped wafers were
all coated with a 1 cc solution of α-aminopropyltriethyoxy silane
in 95/5 methanol/water which was dispensed on the wafer and spun
30 secs. at 5000 R.P.M. The coatings were dried at 30 min.x135°C
followed by curing 30 min.x350°C or 15 minx400°C. After curing
and cooling, the wafers were placed in either boiling water for
two hours or in a pressure pot at 15 psi steam at 230°F for two
hours. After this exposure to moisture the wafers were dried and
scribed in a crosshatch pattern. Mylar adhesive tape was placed
over the scribed areas, smoothed out and then removed with a
quick jerk. The films were examined for adhesion loss by the use
of a microscope.

RESULTS AND DISCUSSIONS
     Polyimides are prepared from the polycondensation reaction be-
tween an aromatic dianhydride and oxydianiline (Fig. 1). In this
paper, we will use the above polymer as typical of all PI coatings. All experimental results quoted here were obtained with
this system.
     For a polyimide to be used as a coating, it must be applied
in a liquid form. Since polyimides are not soluble in most common solvents, (Table I)they are usually applied in the form of
their precursors, polyamic acids. The polyamic acid is soluble
in strongly basic solvents such as NMP, DMAC, etc. (solubility
Table II). The solution of the polyamic acid is applied to the
substrate and thermally cured to form the polyimide.

TABLE I   -   POLYIMIDE SOLVENT RESISTANCE

SOLVENT	EFFECT OF 24 HR. IMMERSION
Acetone	No Attack
Carbon Tetrachloride	No Attack
Cresol	No Attack
Ethyl Acetate	No Attack
Ethyl Alcohol	No Attack
Hexane	No Attack
Petroleum Naphtha	No Attack
Xylene	No Attack
Dimethyl Formamide	No Attack
Antimony Trichloride	Dissolves

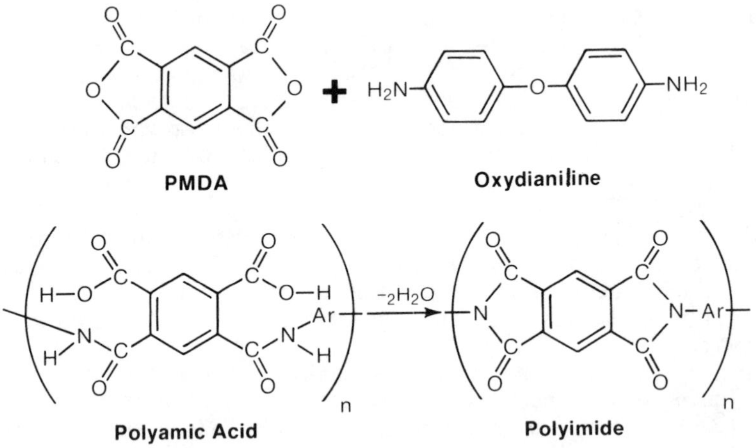

Figure 1.   Polycondensation   reaction   between   an   aromatic   dianhydride   and
oxydianiline.

TABLE II   -   SOLUBILITY OF POLYAMIC ACID
POLYAMIC ACID IS SOLUBLE IN: (Polyimide is not soluble in the
  same solvents)
DMF   DIMETHYL FORMAMIDE
DMAC   DIMETHYL ACETAMIDE
DMSO   DIMETHYL SULFOXIDE
NMP   N-METHYLPYROLLIDONE

The following are the typical processing steps involved in
applying polyimide coatings to a wafer.
When positive photoresist is used:
1) Priming the wafer: To obtain good adhesion between the wafer
and PI coating, a dilute solution of an organosilane is applied
to the wafer and spun to dryness. 2) Applying PI coating: Polyamic acid solution is spun onto the wafer. 3) B-stage the film:
The coating must be prebaked to remove solvent and in the process
the polyamic acid coating is partially imidized. The degree of
this bake determines the conditions for etching and stripping.
4) Photoresist: A solution of photoresist is spun on top of the
PI coating. If a negative resist is used, the processing steps
are described in the following paragraph. 5) Align and expose.
6) Etching and developing: The etching of PI coating and developing of the photoresist can be accomplished in one step with a
dilute aqueous alkaline solution. Typical etchants are dilute
solutions of NaOH, KOH, tetra-alkyl ammonium hydroxide, etc.
7) Neutralization and Rinse: It is imperative that alkaline
metal ions are totally removed from the system. This can be
accomplished by an acetic acid solution wash followed by rinsing
with deionized water 8) Stripping of Photoresist: The photoresist
can be stripped with solvents such as acetone, isopropanol, butyl
acetate, etc. 9) Final Cure: The polyimde coating is given a
bake of 300°C or higher to complete the imidization to obtain
the full film properties.10) Oxygen plasma or chemically and phy-
sically treat the surface to improve adhesion for the next level
metal.   11) Metallization: Deposit aluminum or Al/Cu Alloy.
Negative Photoresist: Step 1, 2, 3, 4, 5 are the same as positive.
6) Develop the image with commercial negative photoresist developer 7) Rinse and bake to harden the photoresist typically at
135°C for 15 minutes. 8) Etch the polyamic acid coating with the
same alkaline etchants described above. 9) Neutralization and
rinse. 10) Stripping the photoresist.
    It is necessary to prebake the PI film to 200°C to improve its
resistance towards negative photoresist with a commercial stripper. After baking, remove the photoresist with a commercial
stripper which is usually composed of phenol, strong mineral
acids and solvents. 11) Neutralization and rinse. 12) final
cure. Typical schedules are 30 min. at 350°C or 15 min. at 400°C.
(10)Plasma,chemically(etching) or physically(roughening) treat
the polyimide surface to improve adhesion for next level metal.
11) Metallization
    To better understand the variables that are important to the
processing as well as final performance of the coating, we stud-

ied the three areas that we regard as most critical. These are:
adhesion,cure cycle, and thermostability.
ADHESION - If a polyamic acid solution is spun onto a wafer (Si
or SiO$_2$ surface) the coating can be readily stripped. Excellent
adhesion, however, can be obtained if the wafer surface is first
primed with an aluminum alchoholate, a collodial alumina, or an
organosilane. All are found to be effective in promoting adhesion. Because of their ease of application, commercial availabil-
ity, and effectiveness at very low concentrations, we chose to
concentrate our study on the Organosilanes. Among all the effective Organosilanes to date studied, we chose ∝-amino propyltrieth-
oxysilane for more detailed investigation. Silanes containing
diamine or triamine such as n-beta amino ethylamino propyltri-
methoxysilane are at least equally effective. Results are shown
in Table III.

TABLE III- 1.0-0.1% - AMINOSILANE (b.w. Boiling Water
                                    p.c. Pressure Cooker)

Wafers	15 mil, 3" diameter
	p-type, boron doped
	1,1,1 polished, any resistivity

PI Control	1 min.b.w.:	0% Peel	
	5 min.b.w.:	50% Peel	
	10 min.b.w.:	88% Peel	
	30 min.b.w.:	100% Peel	

Air Dry Silane			
	1 min.b.w.:	0% Peel	
	10 min.b.w.:	0% Peel	
	60 min.b.w.:	0% Peel	
	30 min.p.c.:	0% Peel	(250°F, 15 psi)
	120 min.p.c.:	0% Peel	(250°F, 15 psi)

10 Min. silane bake at 130°C			
	60 min.b.w.:	0% Peel	
	120 min.p.c.:	0% Peel	(250°F, 15 psi)

THE MECHANISM BY WHICH ORGANOSILANES PROMOTE ADHESION IS
POSTULATED

$$H_2N-CH_2-CH_2-CH_2-\underset{\underset{OCH_2-CH_3}{|}}{\overset{\overset{CH_2CH_3}{|}}{\overset{O}{|}}}{Si}-OCH_2-CH_3 \xrightarrow{\text{hydrolysis}} H_2N-CH_2-CH_2-CH_2-\underset{\underset{OH}{|}}{\overset{\overset{OH}{|}}{Si}}-OH$$

Polymerization
$$\longrightarrow$$

$$\underset{}{Si}-O-\underset{\underset{O}{\underset{|}{}}}{\overset{\overset{(CH_2)_3}{|}}{\overset{\overset{NH_2}{|}}{}}}{Si}-O-\underset{}{Si}-O-$$

SiO$_2$ Surface
/////////

Our proposed mechanism contains the following steps: 1) Hydrolysis of silane. 2) Polymerization of the silane. 3)Condensation of the residual ethyoxy group of the polymer with H_2O that is strongly bonded to SiO_2surface. 4) The formation of ionic bonding between amine and acid form the polyamic acid. 5) At high temperature such as that experienced by the wafer in the typical process, the organic segment of the silane is pyrolyzed.

Experimental evidence in support of the above theory is as follows: 1) Step 1 implies that moisture is necessary for this reaction to take place. Results in Table IV show that adhesion is poor when water is not present in the formulation.

TABLE IV - EFFECT OF MOISTURE ON THE ADHESION OF THE SILANE PRIMER WAFER 0.1% SILANE SOLUTION

Wafer	Applied from	Adhesion in b.w. (30 min.)
3" P-type	Methylene Chloride	100% Peel
same	Methanol	100% Peel
Same	Methanol/Water (95/5)	0% Peel

2) Another indication of hydrolysis is shown by the I.R. spectrum Figure II. The change in $-CH_3$ absorption at 2980 cm^{-1} at room temperature was followed with a F.T.I.R. employing time lapse technique(7). 3) Functional groups that will interact with acid must be present to obtain maximum adhesion. This conclusion is supported by the fact that epoxy and amine containing silanes are all excellent adhesion promotors as shown in Table V.

TABLE V - ADHESION PROMOTING SILANE COUPLING AGENTS

$NH_2CH_2CH_2CH_2Si(OC_2H_5)_3$ 0% Peel

$NH_2CH_2CH_2NHCH_2CH_2CH_2Si(OCH_3)_3$ 0% Peel

$-CH_2CH_2Si(OCH_3)_3$ "

$CH_2-CHCH_2OCH_2CH_2CH_2Si(OCH_3)_3$ "

4) When a silane coated wafer was baked at 400°C and examined by ESCA for surface elements it showed almost a total loss of carbon and amine nitrogen. We view this, therefore, as evidence that organic silane will not survive the bake and SiO_2 will remain. Results are presented in Table VI.

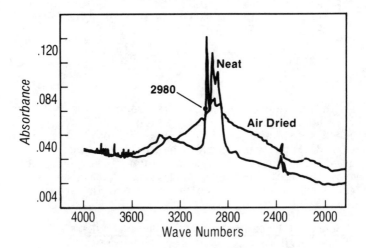

Figure 2.   IR spectrum of δ-aminosilane.

TABLE VI - ESCA RESULTS OF SILANE COATING ON WAFER ELEMENTS

	C	N*	N	Si	O
Room T	6.19	1.17	2.15	1.00	1.94
135°C	6.37	0.29	0.53	1.00	1.83
400°C	0.83	0.28	0.0	1.00	1.96
Blank	0.70			1.00	1.50

N-EB $\cong$ 399 ev Amine Nitrogen
N*   - High E_B Nitrogen Species

We further postulate that although the organic segment of silane was thermally decomposed during the bake, it has however served the purpose of bringing the PI coating into intimate contact with the silicon oxide surface. It can be viewed that the true mechanism for the adhesion enhancement is the formation of interlocking networks of polyimide molecule and SiO_2. Further evidence to support our hypothesis is:
°When amino silane primed wafers were baked at 400°C followed by applying polyamic acid, silane had lost its effectiveness as an adhesion promoter. Results are shown in Table VII. On the other hand when polyamic acid solution was spun onto a wafer that was first primed with silane then coated with polyamic acid solution and baked at 400°C, excellent adhesion was retained.

TABLE VII-EFFECT OF BAKING ON THE ADHESION OF PI TO THE SILANE PRIMED WAFER

Wafers	Primed With Silane	480°C Bake (min.)	Adhesion in b.w. (2 hrs.)
3"P-type	Yes	0	0% Peel
"	Yes	30	100% Peel
"	No	0	100% Peel
"	No	30	100% Peel

All wafers topcoated with PI-2545, 5000 R.P.M. for 60 seconds

CURE CYCLE

Probably the single most important variable in the processing of PI coatings on wafers is the cure cycle. During the heating of the film, solvent evaporation takes place simultaneously with imide formation. The amount of solvent remaining after bake and the degree of imidization will determine the solubility and solvent resistance of the film. When a very thin film is used such as on the wafer (0.5-5 microns) the amount of residual N-methyl pyrrolidone (NMP) was expected to be minimal after baking at 130°-150°C. To verify this, we prepared a 1-mil film from an NMP solution of polyamic acid baked at 150°C for 15 minutes and determined the amount of residual NMP by mass spectrometry. The residual solvent measured only 1.9%. The solubility characteristic of the PI film on wafer is therefore determined primarily by the degree of imidization.

When increasing number of amic acid groups are thermally dehydrated to form imide, the PI film becomes increasingly insoluble. During the process, a condition must be defined so that the film is baked sufficiently to withstand the stripping solvent but not too advanced to cause difficulty during etching. Thus, information on the rate of cure is helpful in defining the processing conditions.

A convenient way to follow the rate of imidization is by monitoring the change of IR absorption. The I.R. spectrum of a polyamic acid differs significantly from that of the corresponding polyimide. Absorption of n-imide at 725 cm^{-1} or 1776 cm^{-1} can be used. We choose to measure the rate of immidization by following the increase in absorption at 1776 cm^{-1} vs time at a given temperature. These results are shown in Figure III.

By using a Fourier Transform I.R. including the computer supported dispersion instruments, spectral substraction can effectively cancel out unchanging bands and expose only those participating in the reaction. Likewise, the spectrum of the substrate (e.g., the wafer) can also be cancelled out. We believe that data obtained this way has increased the sensitivity of this method.

As the results in Fig. IV show that cyclization proceeds at a very slow rate at 135°C but quite rapidly at 180°C, the log of concentration of polyamic acid vs time gives two straight lines which intercept at 90% conversion Fig. V. This can be interpreted to mean that imidization reactions can be divided into rapid and slow, first-order ring closure steps. These results are in general agreement with that reported by Kruez (8).

He also reported the activation energy of 26±3 K cal/mole for the fast reaction and 23±7 K cal/mol for the slow cyclization. The slower rate appears to be due to the entropy of activation (-10 e.u. for the fast reaction and -24 e.u. for the slow reaction). The large increase in the steric factor can be explained by the fact that the polymeric chains become increasingly rigid as a result of cyclization. It is reasonable to expect at a point near 90% imidization, the polymer chain has become greatly rigidized and aligning the reacting groups to the proper position becomes very difficult. It must be noted that the use of I.R. spectrometry including the use of F.T.I.R. will not be able to detect the last few percent (3-5%) for the imidization. This is evident by the fact that film baked at 200°C for 30 minutes showed no further detectable change in I.R. spectra with additional heating. Dissipation factor measurements of the same films, however, showed a continued drop until 340°C. Fig. VI shows the results reported by Gregoritsch.(9)

THERMOSTABILITY - The thermal life of a particular material cannot be described in simple numbers because each particular application has its own criteria for failure. For example, dielectric breakdown is the mode of failure for an electrical

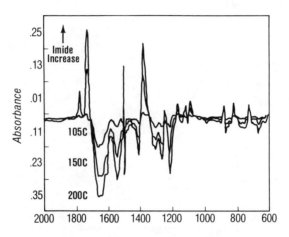

Figure 3. Polyimide cure study.

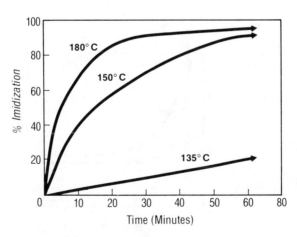

Figure 4. Polyimide cyclization.

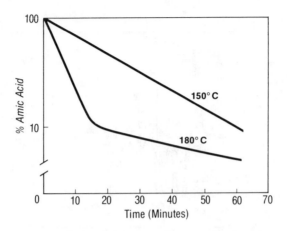

Figure 5.   Reaction rates of imidization.

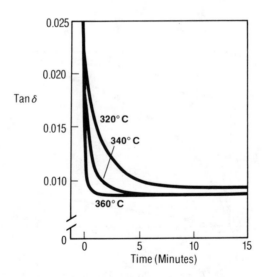

Figure 6.   Tan δ vs. time (on hot plate) for films.

insulation in motors but loss of elongation is the cause for
flexible cable failures. The thermal breakdown of polyimide
is known to proceed according to at least three different
mechanisms. The deterioration in air can be attributed to a
radical-initiated oxidation. At lower temperature
and high humidity, the predominating reaction is hydrolysis.
In the absence of air and moisture, the polyimide degrades
through a pyrolytic reaction. Since radical-initiated oxidation
can be catalyzed by the presence of transition metals, the nature of the substrate will have a profound effect on the thermal
life of the film in air.

   Thermogravimetric analysis, though not necessarily indicative of all high temperature properties, is a convenient way to
indicate the degree of thermal reaction occurring at a particular temperature. The thermal stability of polyimides has been
the subject of many studies (10). Our results obtained from
free films of 1-2 mils are shown in Fig. VII and are in agreement with Heacock and Berr(11). Since 1-2 mil films are not
used here, we decided to study the thermal degradation of

polyimide coatings on the wafer in air and nitrogen to simulate
more realistic conditions. This work was aided by the use of
the Du Pont 1090 Thermalanalyzer with its greatly expanded
sensitivity and versatility.

   The weight loss of PI coatings thicknesses of 6-12 microns on
wafers and on aluminum (sputtered on) at 450°C and 500°C in
air and in N_2 were determined. Weight loss data on the same
films stripped off from the wafers were used as controls. Fig.
VIII lists the experimental results. These results show that
the thermal life of the polyimide film is not affected by the
presence of SiO_2. The aluminum coated PI actually showed a
lower weight loss in air assuming all the aluminum has been
oxidized to aluminum oxide. One possible explanation is that
at 500°C, the oxygen in air reacts with aluminum preferentially
thus reducing the concentration available to attack the PI film.

   As expected the weight loss in N_2 is much less than in air.
This is consistant with the expectation that the pyrolytic reaction of polyimide should have a much higher activation energy
than the radical-initiated oxidation. It is noted that the loss
on wafer at 500°C in air appear to be much more than those of
thick films obtained previously.

   Thermostability requirement for microelectronic applications
basically involves only the thermo exposure during processing.
Since the devices are not expected to operate at anywhere near
the processing temperature. At 400°C in air, even with very
thin films polyimide do not show any sign of degradation within
the time (30-60 min) processing take place. We, therefore,
conclude that fully aromatic polyimide is thermally sufficient
for this application.

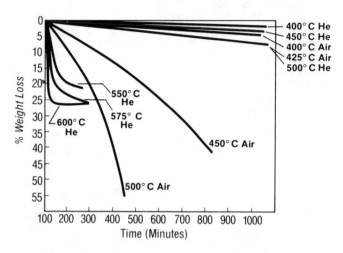

Figure 7.   Isothermal weight loss (polyimide 1 mil).

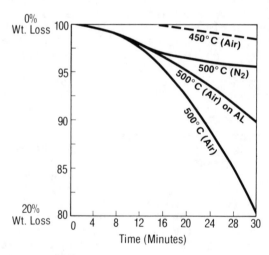

Figure 8.   Weight loss of TGA polyimide (12μm).

CONCLUSION
    Polyimide coatings that are synthesised from Oxydianiline
and Pyromellitic dianhydride can be used as dielectric/passiva-
tine coatings in micro electronic devices. Successful processes have been developed by using either negative or positive
resists. Organo silanes are effective as ahdesion promoters
provided they are partially hydrolysed. ESCA results showed
the silane pyrolysed during processing. Study of the rate of
imidization confirms the existance of a fast and slow two step
reaction. A fast reaction for the first 90% conversion followed
by a much slower reaction for the remaining 10% imidization.
Thermal stability study indicated that polyimide should be
sufficiently thermally stable to withstand the typical process-
ind temperature in the wafer fabrication process.

Abstract
    Conditions have been defined for applying polyimide coatings
onto the silicon wafer as passivation and/or dielectric. Processing variables studied included the critical areas of adhesion, cure cycle and thermostability. Aminosilane was shown to
be effective adhesion promoter. The rate of imidization was
followed by F.T.I.R. employing time lapse technique.

Literature Cited
1. S.A.Evans, et al. IEEE Trans Electron Devices,ED-24,196(1977)
2. L.B.Rothman JECS 127 10, 2216 (1980)
3. K.Mukai, A. Saiki, K. Yamonaka, S.Harada and S. Shoji,
      IEEE Journal of Solid State Circuits, SC-13,No4,462(1978)
4. M. Gold Electronics News 11/6/78
5. G.Samulson, Org. Coatings & Plastics Chemistry
      Vol. 43, 446, (1980)
6. T. Herdon.   Private Communication
7. J.A.Hartshorn. Applied Spectroscopy, 33, 2 (1979)
8. J.A.Kruez, A. Endrey, F.P.Gay and C.E. Scroog
      J. of Poly. Science 4,2067(1966)
9. Gregoritsch, 14th Annual Proceeding of the reliability
      Physics, P.228 (1976)
10. J. SPE Transaction 105, April, 1965
11. J.E.Heacock and C.E.Berr. SPE Transactions 5, 2(1965)

RECEIVED October 29, 1981.

Development of Polyimide Isoindoloquinazoline-dione in Multilevel Interconnections for Large-Scale Integration (LSI)

ATSUSHI SAIKI, KIICHIRO MUKAI, and SEIKI HARADA

Hitachi, Ltd., Central Research Laboratory, Kokubunji 185, Japan

YASUO MIYADERA

Hitachi Chemical Co., Ltd., Shimodate Research Laboratory, Shimodate 308, Japan

For LSI's, higher packing density requirements are very stringent and multilevel interconnections are indispensable in satisfying these requirements. Planar Metallization with Polymer (PMP) technology has been developed to realize these highly packed multilevel interconnections. In this technology, a polymer, instead of more commonly used chemical vapor deposited silicon dioxide, is employed as an insulating layer(1). A new polyimide type resin, polyimide isoindoloquinazolinedione (PIQ R) has been synthesized for this purpose.

In conventional two-level metallization, chemical vapor deposited silicon dioxide is used as an insulating layer, as shown in Figure 1(a). In this structure, there are many steps at metal (aluminum is most often used) layer edges and the through-holes in insulating layers. Since the coverage of both metal and insulating layers at the steps is very poor, defect density is large and fine patterning of the second aluminum layer is difficult.

In contrast, the PMP structure, shown in Figure 1(b), is an ideal planar structure. PIQ films are formed by thermal curing after spinning PIQ prepolymer solution on a silicon wafer. Since the fluid properties of PIQ provide ideal wafer flatness, the PIQ insulation permits excellent step coverage in overlapping metal conductors, no matter how many steps are formed by previous metallization. This technology offers higher reliability and increased yield over conventional techniques. In order to realize PIQ application to semiconductor devices, synthesis of an extremely high heat resistant and pure PIQ is essential. Therefore, in this paper, the properties of PIQ material, especially heat resistance and purity, are described. Application of PIQ to semiconductor devices is also described.

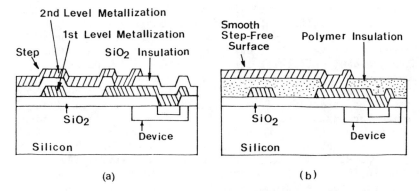

Figure 1. Comparison of two-level metallization structure by (a) conventional method and (b) PMP technology.

Experiments

Synthesis of PIQ. Very high heat resistance is required in order for a polymer film to be used as an insulator. This is because several heat treatments over 400 C are necessary in LSI interconnection and assembly processes. An aromatic polyimide (I), a reaction product of aromatic diamine and acid dianhydride, is one of the most heat resistant polymeric materials:

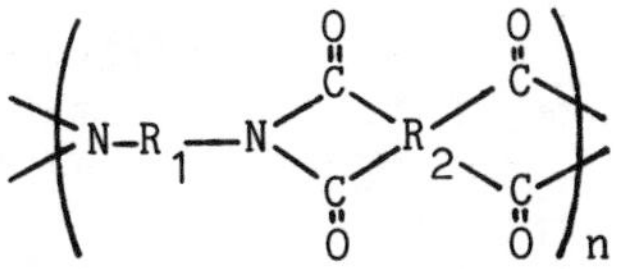

(I)

where R_1 and R_2 are aromatic rings. A new ladder structure, isoindoloquinazolinedione (II), was created and introduced to a polyimide chain to satisfy heat resistance requirements:

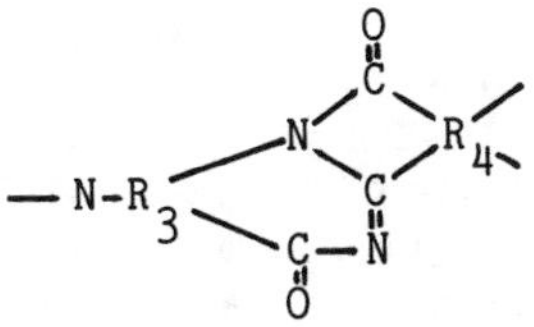

(II)

where R_3 and R_4 are aromatic rings. In order to realize this structure, new aromatic diaminocarbonamide(III), was synthesized:

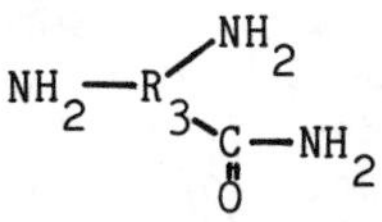

(III)

PIQ is synthesized from this aromatic diaminocarbonamide, aromatic diamine and two aromatic dianhydrides. As a result, PIQ consists of isoindoloquinazolinedione and an ordinary polyimide, that is polyimide isoindoloquinazolinedione (IV):

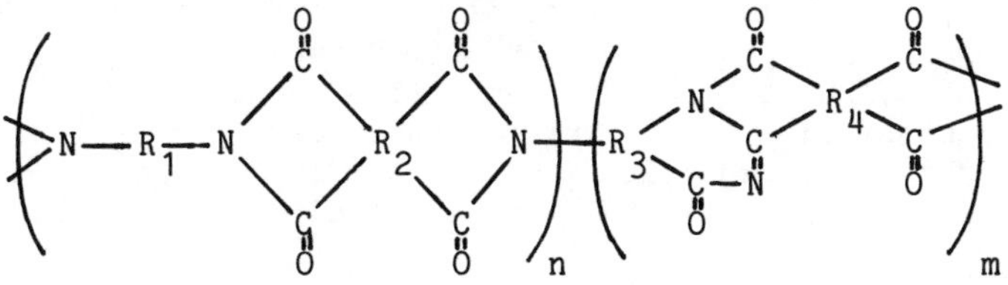

(IV)

This provides higher heat resistance, as will be described
later.

PIQ Film Fabrication. Fabrication of the PIQ film is
as follows. PIQ prepolymer solution is dispensed and then
spin coated on a silicon wafer followed by thermal curing.
A spin speed of from 3000 to 5000 rpm and an 1.1 Pa s (11
poise) solution makes a 1.5 to 2.0 um thick cured film.
Curing is first performed at 200 C in air for 1 hour and
then at 350 C in a nitrogen atmosphere for 30 minutes.

Heat Resistance Evaluation. PIQ cured films were kept
at high temperatures, and then change in film thickness,
visual light absorption, infrared absorption and film weight
were measured. A Taylor Hobson Talystep was used for film
thickness measurement. A Hitachi 124 spectrophotometer and
Hitachi-Parkin Elmer 225 infrared spectrophotometer were
used for visual light absorption and IR absorption. Film
weight were measured using Mettler type M5 microbalance.

DuPont polyimide, Pyer-ML RC5057 was used for
comparison. Pyer-ML varnish was diluted to 1.1Pa s (11
poise) using N-methyl-2-pyrrolidone and spin coated on
silicon wafers and thermally cured.

Analyses of Water Content. The water content of the
PIQ starting materials was analyzed. The water content of
amines was measured using a DuPont 321A moisture meter and
those of the solvents were measured by Karl Fischer's
reagent method. The water content of acid dianhydrides was
measured by titrating the free acid.

Impurities in PIQ and Starting Materials. Metallic
impurities contained in both PIQ and starting materials were
analyzed using a Hitachi 308 atomic absorption spectrometer.

A specially designed transistor was used to investigate
the influence of the impurities in PIQ on transistor
characteristics. The electrodes of this transistor do not
completely cover the contact holes of the emitter and base.
This structure is very sensitive to contamination.

Results and Discussions

Heat Resistance. The heat resistance of PIQ was
determined by measurement of film thickness changes. PIQ
prepolymer solution was coated on silicon wafers and
thermally cured. The final cure temperature was 350 C in a
nitrogen atmosphere. Then, samples of the films were heated
at temperatures ranging from 200 C to 450 C for 5 hours in
air. After this, the thickness of each film sample was
measured and compared with its initial thickness. The
results are shown in Figure 2. The results for ordinary
polyimide, DuPont Pyer-ML RC5057, are also indicated in this
Figure.

From the figure, it is clear that the PIQ film
maintains its initial thickness even after heating at 450 C

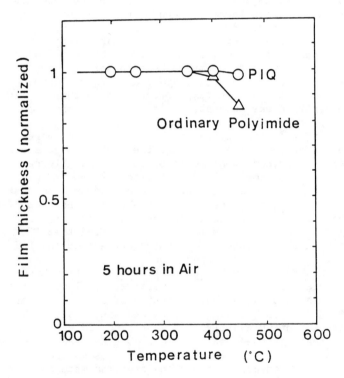

Figure 2. Heat resistance of PIQ evaluated by film thickness change.

in air for 5 hours. On the other hand, the ordinary
polyimide loses 15 to 20% of its initial thickness after the
same thermal treatment. This difference in heat resistance
is obviously due to the introduction of the new polymer
structure.

Thermal Lifetime. The absorption of short wave length
visual light transmitted through a PIQ film increases when
the film is heated. The thermal lifetime of PIQ film was
defined as the time until the absorbance of 450 nm wave
length light transmitted through a 1 um thick PIQ film
reached 0.5. Films were formed on quartz plates and heated
at high temperatures in a nitrogen atmosphere until the
absorbance became 0.5.

The results are shown in Figure 3. The lifetime
linearly changes inversely with the temperature. Some
examples of the time-temperature relationship for typical
treatment processes are also indicated in this Figure. It
can be seen that the time-temperature relationship of PIQ is
far greater than any encountered in LSI interconnection and
assembly procsses.

The infrared absorption spectra of PIQ films were
measured before and after the lifetime tests, to determine
the degree of degradation caused by the tests. Figure 4
shows the results for a PIQ film treated at 450 C for 5
hours in nitrogen. No degradation in any absorption peak
can be seen in this Figure. For example, the peaks at 1780,
1730 and 730 cm^{-1}, for the imide carbonyl bond, are
unchanged. From these results, it is verified that PIQ has
sufficient heat resistance for heat treatment during LSI
fabrication and long term operation.

Influence of Water on Heat Resistance. It is known
that the viscosity of polyimide prepolymer (polyamic acid
solution) is reduced by the absorption of a·small amount of
water(2). This is because the absorbed water reduces the
molecular weight by hydrolysis. Therefore, it seems
probable that absorbed water would reduce the heat
resistance of cured PIQ film. For that reason, the
influence of water contained in the starting materials on
the heat resistance of cured PIQ film was investigated.

The water content in the starting materials was
measured as follows. The water content of amines and
solvents were measured using the DuPont 321A moisture meter
and Karl Fischer's reagent methods, respectively. Since the
water contained in acid dianhydride is considered to convert
it to free acid,

$$O\diagdown \!\!\!\begin{array}{c} CO \\ CO \end{array}\!\!\!\diagup R \diagdown\!\!\!\begin{array}{c} CO \\ CO \end{array}\!\!\!\diagup O \quad \xrightarrow{2H_2O} \quad \begin{array}{c} HOOC \\ HOOC \end{array}\!\!\!\diagup R \diagdown\!\!\!\begin{array}{c} COOH \\ COOH \end{array}$$

the water content was measured by titrating the free acid.

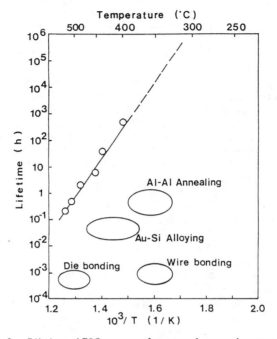

Figure 3. Lifetime of PIQ compared to several processing temperatures.

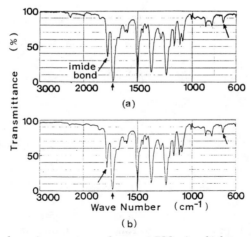

Figure 4. IR absorption spectrum change of PIQ after high-temperature heating (a) as prepared and (b) after heating in nitrogen at 450°C for 5 h.

The results are shown in Table 1. The table shows that the greatest part of the water in the starting materials is present in the acid dianhydrides. This was verified by experimental results. That is, water content increases with time and that increase strongly depends on ambient humidity, as shown in Figure 5. Therefore, it is very important to keep the materials dry.

In order to reduce the water content, dehydration of the starting materials were carried out as follows. Acid dianhydrides were recrystallized in acetic anhydride and dried by infrared lamp. Amines were recrystallized in butyl alcohol and dried. Solvents were distilled under reduced pressure. The water content of the dehydrated materials is also given in Table 1. A remarkable reduction in water content was achieved.

The influence of water on heat resistance of cured PIQ film was examined as follows. PIQ prepolymer solutions were synthesized using moist acid dianhydrides and PIQ films were formed from these solutions. Next, heat treatment was carried out on the films at 450 C in air and their weight residues were measured. The results are shown in Figure 6. It was found that even a small amount of water greatly decreased the heat resistance of PIQ.

The influence of water on heat resistance is considered to be as follows. Moist acid dianhydrides include free acid, as mentioned before. When PIQ is thermally cured, free acids are converted into corresponding dianhydride monomers and evaporate.

Since the free acids do not take part in polymerization, they prevent high polymer formation and reduce heat resistance.

High Purity PIQ Synthesis. PIQ must not degrade device characteristics. The influence of PIQ application on transistor characteristics was evaluated using an npn test

Table 1. Water Content in PIQ Starting Materials

MATERIALS		WATER CONTENT (wt%)	
		Raw	Dehydrated
Amines	AM	0.01	0.00
	AMC	0.11	0.08
Acid	DA-1	3.93	0.01
Dianhydrides	DA-2	0.05	0.00
Solvents	NMP	0.06	0.00
	DMA	0.06	0.00

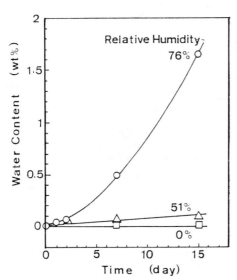

Figure 5. *Influence of relative humidity on water content in acid dianhydride.*

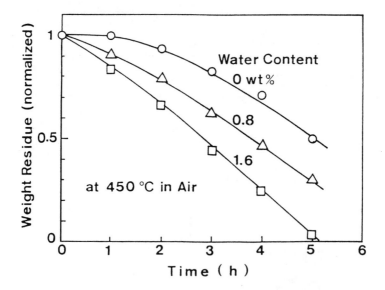

Figure 6. Influence of water content in acid dianhydride on PIQ heat resistance.

transistor. The aluminum electrodes of this transistor do
not completely cover the contact holes of the emitter and
collector thus exposing non-passivated areas. After the PIQ
film was formed the test transistors were annealed at
temperatures of from 100 to 500 C for 30 minutes, and the
current gain, h_{FE}, was measured. The results are shown in
Figure 7. Each curve corresponds to a differently
synthesized PIQ. Some PIQ's did not degrade the transistor
characteristics even after annealing at high temperatures,
but others degraded them drastically. Since h_{FE} is very
sensitive to metallic impurity contamination, especially
sodium ions, the sodium ion content in PIQ was analized by
atomic absorption spectroscopy. The results of this
analysis are also indicated in Figure 7. With a higher
sodium content, h_{FE} drastically decreased as temperature
increased. On the other hand, no change was observed for
low sodium content up 500 C annealing.
    Sodium ion content in the raw materials and PIQ was
analyzed. The results are shown in Table 2. It can be seen
that sodium ions are present primarily in the amine and
aminocarbonamide.
    The reason for so many sodium ions in diamines can be
explained as follows. Diamine is prepared by reduction of
dinitro-compounds(V). The diamine then reacts with
hydrochloric acid:

$$O_2N-R-NO_2 \xrightarrow{\text{reduction}} H_2N-R-NH_2 \xrightarrow{\text{HCl}}$$

(V)

$$HCl-H_2N-R-NH_2-HCl \xrightarrow{\text{NaCl}} H_2N-R-NH_2 + NaCl$$

(VI)

forming hydrochloride(VI) to separate organic impurities.
Then hydrochloride is neutralized by sodium hydroxide to
form diamine and sodium chloride. Therefore, many sodium
ions from sodium hydroxide and sodium chloride are contained
in the raw amine materials. This also applies to
diaminocarbonamide.
    The raw materials were purified in a manner similar to
dehydration. The sodium content in the purified materials
is also given in Table 2. It should be noted that the PIQ
synthesized from raw materials had a sodium content of as
much as 54 ppm, while the PIQ from purified materials
contained less than 1 ppm.
    Further experiments were carried out on npn and pnp
type transistors, and results are shown in Figure 8. In
that Figure, the h_{FE} change ratio after 500 C annealing is
shown as a function of sodium content. It was found that
PIQ, which contained less than 3 ppm sodium, did not change

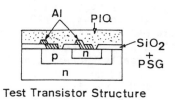

Test Transistor Structure

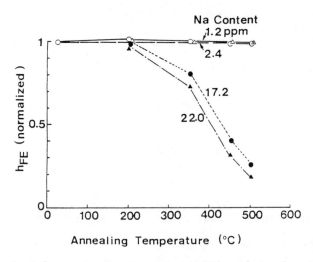

Figure 7.   Influence of sodium ion content in PIQ on device characteristics.

Table 2.   Sodium Content in Starting Materials
and PIQ Varnish

MATERIALS		SODIUM CONTENT (ppm)	
		Raw	Purified
Amines	AM	185.0	4.5
	AMC	792.0	1.2
Acid	DA-1	2.0	0.6
Dianhydrides	DA-2	1.9	1.1
Solvents	NMP	0.56	0.07
	DMA	0.55	0.07
PIQ Varnish		54.0	0.51

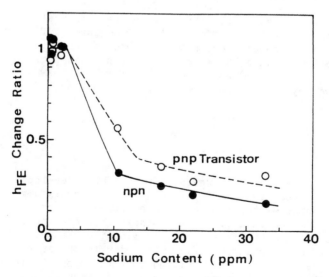

Figure 8.   Influence of sodium ions in PIQ on h_{FE} stability.

the characteristics of either npn or pnp transistors.
Therefore, material purification is very effective and
important for realizing PIQ with the high purity absolutely
necessary for semiconductor devices. PIQ can now be
produced, on the line, with a sodium content of less than
0.5 ppm.

LSI Application

The most effective use of PIQ in semiconductors is as
an insulating layer for multilevel interconnections. After
the first metallization is formed, a PIQ prepolymer solution
is spin coated on the wafer and then thermally cured.
Curing is first performed at 200 C in air and then at 350 C
in a nitrogen atmosphere. Through-holes are etched using a
negative photoresist as the etching mask and an etchant
containing hydrazine hydrate and ethylenediamine. Next, the
second level metallization is formed. PIQ film is then
formed on the second level metallization. This acts as a
passivation film.

An LSI produced by two-level metallization with PIQ
insulation is shown in Figure 9. This is an exposure
controler for an automatic exposure camera. Several hundred
7x7 um^2 through-holes are formed in a 2 um thick PIQ
insulator and connect the first level and second level
metallizations. This LSI is molded in a plastic package and
has good reliability due to the use of PIQ, not only as an
insulating layer but also as a protective coating(4).

Conclusions

A new polyimide type resin, PIQ, has been developed for
use in higher packing density LSI's. The extremely high
heat resistance of the PIQ has been achieved by introducing
a new ladder structure, isoindoloquinazolinedione, into the
polyimide chain. It is necessary to minimize the water
content of the starting materials in order to obtain this
high heat resistance. It was found that the upper limit of
Na ion content in PIQ which did not degrade transistor
characteristics was 3 ppm. Multilevel metallization
technology using PIQ insulation is already in a production
stage and LSI's utilizing this technology are being
successfully produced, with high reliability.

4.3 x 4.5mm²

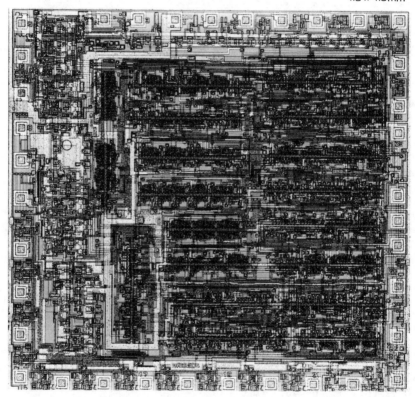

Figure 9. An LSI employing two-level metallization and PIQ insulation.

Acknowledgements

The authors wish to thank Dr. Toru Abe, President of
the Iruma Electric Co., Ltd.; Dr. G. Kamoshita, President
of the Hitachi Microcomputor Enginnering Co., Ltd.; Dr. T.
Muroi, Director of the Hitachi Plant Engineering Co., Ltd.;
and Dr. K. Sato, General Manager of the Central Research
Laboratory, Hitachi, Ltd. for their helpful advice and
continued encouragement. Mr. T. Takagi and Mr. T. Okubo
contributed greatly to the fabrication and evaluation of the
test transistors. Mr. K. Yamanaka and Mr. K. Kuga are
also thanked for carrying out the visible radiation
absorption spectroscopy and atomic absorption spectroscopy,
respectively.

Literature Cited

1) K. Sato, S. Harada, A. Saiki, T. Kimura, T. Okubo
   and K. Mukai, A novel planar multilevel interconnection
   technology utilizing polyimide, IEEE Trans. Parts,
   Hybrid and Packaging, PHP-9, pp.176-180, Sep. 1973.
2) D. J. Parish and B. W. Melvin, Kapton polyimide film
   in rotating machinery, a paper presented at the 6th
   Electrical Insulation Conference (IEEE and NEMA), Sep.
   1965.
3) A. Saiki, S. Harada, T. Okubo, K. Mukai and T.
   Kimura, A new transistor with two-level metal electrodes,
   J. Electrochem. Soc., 124, pp. 1619-1622, Oct. 1977.
4) K. Mukai, A. Saiki, K. Yamanaka, S. Harada and S.
   Shoji, Planar multilevel interconnection technology
   employing a polyimide, IEEE J. Solid-State Circuits,
   SC-13 (4), pp 462-467, Aug. 1978.

RECEIVED October 19, 1981.

Characterizing Polyimide Films for Semiconductor Application

A. M. WILSON[1], D. LAKS[2], and S. M. DAVIS[3]

Texas Instruments, Semiconductor Research and Development Laboratory, Dallas, TX 75265

Polyimide resins as a class of chemical compounds have been available since 1926 ([1]). Their usefulness to the electronics industry, as a plastic insulator which retained electrical and mechanical characteristics at temperatures in excess of 250°C, was recognized early ([2]). Harada and coworkers at Hitachi Central Research Laboratories reported on a Polyisoindolo-quinazolinedione, given the acronym PIQ, which was suitable as an insulator for multilayered interconnections on large scale integrated (LSI) circuit surfaces as well as a final passiviation overcoat film ([3]). R. Rubner and coworkers described a photo-sensitive polyimide which should shorten the semiconductor processing steps ([4]). L. B. Rothman is the first author to evaluate and compare commercially available polyimide resins by testing for characteristics vital to their use in LSI and very LSI (VLSI) circuits ([5]). This paper reports on procedures we have used to characterize and screen polyimides so that we might develop processes which use polyimides as overcoats and insula-tors for single level and multilevel interconnect systems.

Experimental

Chemicals & Materials PI2545 (Type I polyamic Acid) and PI2555 (Type III) were obtained from I.E. Dupont, Marshal Labora-tories, P.O. Box 3886, Philadelphia, Pa. 19146 as 10-12 poise solutions, containing 14 and 20% solids, respectively. PIQ (polyamic acid iso-indroquinazlinedione) resin and coupler were supplied by Hitachi Chemical Co. American, Ltd., 437 Madison Avenue, New York, N.Y. 10022. All other chemicals were regent grade or better. Semiconductor grade 76 mm diameter, 0.45 - 0.55 mm thick silicon wafers with 2 KÅ thermally grown silicon oxide were used as substrates on which to cast all thin polyimide

Current addresses:
   [1] Linear Circuits, Dallas, TX 75265.
   [2] Dallas MOS II, Dallas, TX 75265.
   [3] University of California, Department of Chemistry, Berkeley, CA 94720.

0097-6156/82/0184-0139$05.00/0

films. Resins were used as received or filtered through Balstron
No. 100-12-AAQ filter (0.3 μm average pore size) in a Type 33G
holder. Wool felt filter bag material with average pore size of
one micrometer was obtained from American Felt & Filter Co.,
Newburgh, New York 12550. This material was cut into circular
discs and supported in a 50mm diameter high pressure stainless
steel Millipore filter holder.

Equipment All silicon wafers and thin polyimide films were
weighed with a Mettler micro gram Balance with a standard reading
error of ± 0.020 mg. Polyimide films were cast on commercially
available spin coaters, with point of delivery filtering. Chuck
speeds were calibrated with a stroboscopic tachometer and are in
error by ± 100 rpm or 5%, whichever is larger. Films were annealed in a Thermco Minibrute tube furnace; the active region was
profiled and found to control ± 3°C of the set point. Differential scanning calorimetry (DSC) was done with a Dupont Model 990
Thermal analyzer. An Aminco Thermalanalyzer equipped with a
Marshall Furnance was used for thermal gravimetric analysis (TGA).
Film thicknesses were measured by spectrometry with a Nanospec
Thin film analyzer assuming a refractive index of 1.78 for the
cured film.

Procedures Polyamic Acid films were cast by delivering 1.5
to 2.0 ml of resin to the center of a non-spinning 76 mm dia.
wafer with a positive displacement bellows-type pump: Spinning
then progressed via 1K rpm/sec^2 ramp to the final spin speed for
20 seconds. Films were dried 30 minutes at 90° ± 10°C while held
horizontally in a 25" mercury vacuum. B-stage cures were performed at 205°C for one to two hours in a forced air oven. Unless
noted otherwise, all annealing was done at 350°C in O_2 for 30
minutes.
    Samples to be studied as free films were only b-stage cured,
then the film and wafer were soaked in deionized water overnight.
The film was removed from the wafer by applying a cellophane tape
across an edge of the polymer film and a supporting paper; the
film and wafer were then inverted and an approximate 180° peeling
force applied leaving the film attached to the paper via the
cellophane tape. Permeability studies were performed per a modified ASTM Procedure E96-66 (6). B-staged films were placed over
the mouth of a 30 ml beaker containing 25 ml of dry 30 mesh calcium chloride, folded and affixed by a wire ring of copper. The
mounted sample was redried at 205°C for one hour in a forced air
oven before the final annealing step in the Thermco Minibrute
tube furnance. Hydration was carried out at 23 ± 1°C in a 100%
relative humidity.
    Films to be studied chronogravimetrically for moisture
desorption and absorption were cast on wafers with a 2KÅ thick
oxide substrate film. These substrates were dried at 205° for
two hours, cooled in a desiccator with activated silica gel and

then weighed before casting the polyimide film. Since polyimide films gain weight when stored over-night in the presence of activated silica gel, calcium chloride was used as the dessicant for substrates coated with polyimides.

A 3/8 inch diameter aluminum or titanium-tungsten dot pattern was fabricated on top of the cured polyimide film to make electrical leakage to substrate measurements for pinhole density estimation. An etch decoration technique was used to visually determine pinhole densities in polyimide films. The polyimide film was cast on substrates comprised of a layer of 200 nm thick aluminum on blue colored field oxide with a grid pattern for area computation. Replicate holes were etched in the aluminum by a hot phosphoric acid solution. With the polyimide film removed, a good visual contrast was achieved for pinhole density counting.

Results & Discussion

Coating Characteristics Figure 1 is a typical coating curve for Dupont PI2545 and Hitachi PIQ resins. The materials have approximately the same viscosity and percent solids and the film thickness are, within experimental limits, indistinguishable. In order to be useful as semiconductor insulators, these films must be free of pin holes. Surface cleanliness and the use of adhesion promoters minimize pin holes due to dirt particles and due to loss of adhesion. Surfaces like aluminum oxide formed from oxidative hydration of aluminum tris-acetylacetonate or surface bonded siloxanes with amine terminal groups like γ -glycidoxyl-propyltrimethoxy silane may be utilized to promote adhesion of the polyimides to clean oxides surfaces. Even when such couplers are used, the as-received resin has a pin hole density dependence upon film thickness as shown in figure 2. A felt wool bag filter was used in an attempt to reduce the particle density. However, the defect density actually increased due to the sub one micron particles in the natural fibre, see figure 2. The fiberglass filter, with an inorganic binder and an average 0.3 micron pore supplied by Balstron Company is effective in reducing the particles to less than 0.1 defect/in^2 (zero particles in 24 in^2 of area inspected) in 2 micron thick films.

B-Stage Cure and Annealing Characteristics The cure cycle for polyamic acid resins is complicated by the simultaneous evaporation of the solvents, the closure of the imide ring, the loss of highly bound water and internal polymer chain rearrangements. Due to the high volatility of its primary solvent, 1-methylpyrrolidone, at low temperature, significant weight loss occurs at or below 100°C but continues up to the boiling point of 205°C. A typical TGA curve of Dupont PI2545 in air, which had been cured for 20 minutes at 70°C and 30 minutes at 120°C is shown in figure 3. Three major weight loss regions are observed:

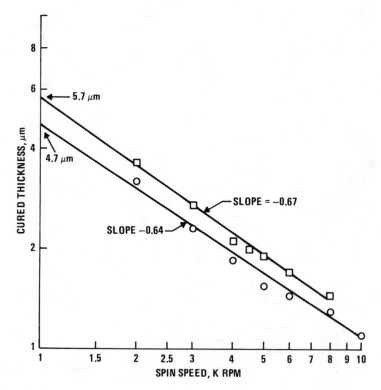

Figure 1. Polyimide coating curves. Key: ○, *Hitachi PIQ, 1100 cps and* □,
DuPont PI2545, 1200 cps.

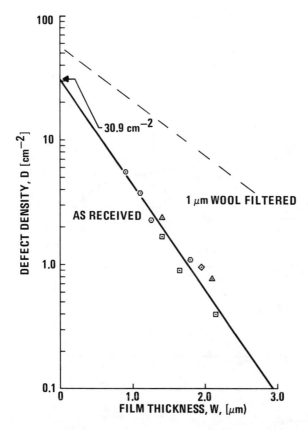

Figure 2. Film thickness dependence of polyimide pin holes. Key: (polyimide type) DuPont Type I—○ and ⊙ , electrical and △, visual; Hitachi PI—□, electrical and ◇, visual; – – –, 1-μm wool filtered and ———, as received.

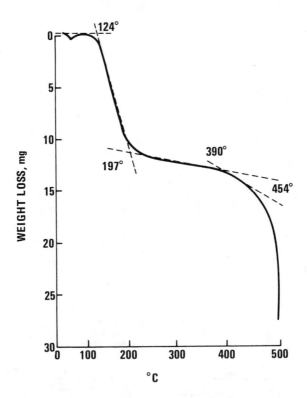

Figure 3.   Thermal gravimetric analysis of DuPont PI2545. Conditions: tempera-ture scan, 4°C/min; precured 20 min at 70°C and 30 min at 120°C; sample weight, 46 mg.

20-25% is lost between 124-197°C, another 5-10% is lost between 350-454° and beyond 454°C the film oxidizes or destructively decomposes. Thermal gravimetric analysis of incompletely anneal-ed Hitachi PIQ and Dupont PI2545 films provides curves with similar regions and little is learned about the curing mechanism from such analysis.

Dupont has studied the rate of imidization by differential IR spectroscopy and found PI2545 type resins to have 95% ring closure after one hour at 180°C (7). It was hoped that the imidization reaction could be illucidated by DSC analysis. A typical curve of duplicate samples run at 1X & 5X Y-axis amplifi-cation is shown in figure 4. The characteristic endothermic peak at 165°C does decrease in specific caloric content as more solvent is removed by vacuum drying at 100°C. Furthermore, the peak is shifted to higher temperatures as samples are progressively cured for longer times at 165°C. However, the peak is only completely removed after the sample is taken to 205°C (8). It appears the energy input is equally distributed between the imidization reaction and solvent evaporation in this temperature range and DSC analysis can be used only as a measure of completeness of imidization. The method used by Dupont is still the best for following the kinetics of the imidization reaction.

The tangent of the capacitance phase angle has been shown to be a measure of the completeness of the anneal (9). Based on this technique, we have data which would predict 30 minutes at 300°C is sufficient to obtain optimum dielectric properties. However, we have found chronogravimetric analysis of polyimide films to be more useful in defining optimum anneal conditions. A typical set of chronogravimetric curves for Dupont PI2545 are shown for different anneal temperatures and atmospheres in figure 5. The relative weight changes are referenced to weight taken on samples cured for two hours at 205°C and cooled for one hour in a desiccator. After standing 24 hours or longer, controls actually gain 3% weight because they are better desiccants than active silica gel. In order to remove this water, temperatures of 350°C for 30 minutes or longer are required. Limiting weight losses of 3.5 and 7% are obtained after 60 minutes at 400°C and 30 minutes at 450°C respectively. This weight loss corresponds to one and two molecules of water removed per polyimide repeat linkage (mer). After rehydrating these samples, all gain 4 to 5% weight. This weight gain corresponds to one and a half water molecules per mer. In the case of the control samples, the PI2545, cured to only 205° and rehydrated, contain approximately seven water molecules per 2 polyimide mers. These waters can easily be accounted for by intra- and inter-molecular hydrogen bonding structures (10). Reactivation by water is not complete once the films are cured for 30 minutes or more at temperatures in excess of 350°C. This is an indication of some intermolecular or further intra-molecular rearrangement which makes intra-molecular water hydrogen bonding less facile.

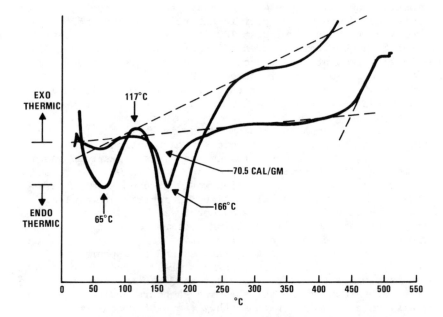

Figure 4.   Differential scanning calorigram of DuPont PI2545.   Duplicate samples run at 1× and 5× amplification of Y-axis.   Conditions: precured 20 min at 70°C and 30 min at 120°C.

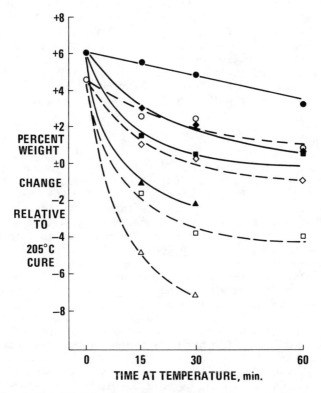

Figure 5.   DuPont PI2545 moisture change vs. anneal time and temperature. Conditions: ○, 350°C N₂; ●, post 113 h 100% RH, 25°C; ◇, 350°C O₂; ◆, post 113 h 100% RH, 25°C; □, 400°C O₂; ■, post 113 h 100% RH, 25°C; △, 450°C N₂; ▲, post 113 h 100% RH, 25°C.

The water permeability data shown in figure 6 shows the optimum anneal temperature to be 350°C for Hitachi PIQ for both O_2 & N_2 atmospheres. Dupont PI2545 and PI2555 optimum anneal temperatures are a function of the atmosphere. Dupont films color and strength are darkened and weakened respectively by the cure in nitrogen (8). Dupont PI2545 is a simple copolymer of pyromellitic acid dianhydride (PMADA) and di-(4,4'diaminophenyl) ether (DAPE), PI2555 is probably also a similar simple copolymer of a single anhydride and a single diamino base; PIQ is a copolymer of PMADA, 3,3',4,4' - benzophenone tetracarboxylic dianhydride (DTDA), DAPE and di-(4,4' diamino-3-carboxylamidephenyl) ether (DACPE). The possibilities exist, either the extra ring created by DACPE provides higher temperature stability to PIQ or the PIQ is synthesized from purer starting materials and has fewer smaller chain length polymers thus displays less carbonization in the 400°C inert atmosphere anneals.

Conclusions

In order to develop polyimide films for semiconductor applications, many chemical, mechanical, thermal and film forming characteristics must be determined in order for the user to develop a viable process for reliable integrated circuits. Dupont PI2545 and Hitachi PIQ are commercially available resins which can be coupled and filtered at the point of use, to provide 1 to 4 mm thick films with less than 0.1 defect/in^2. Differential scanning calorimetry is useful in defining transition temperatures where solvent losses are initiated, however it can not be used to separate the onset of solvent loss from the onset of the internal ring closure which converts polyamic acids to polyimides in the 140 to 205°C temperature range. Thermal gravimetric analysis by scanned technique can not provide accurate measures of stoichiometry of water loss from the polyimide film in temperature range above 300°C. The kinetics of this process are highly hindered and chronogravimetric studies carried out on the minute scale must be used to follow this slow reaction rate. Although free standing polyimide films are highly permeable to water penetration the rate of water transport is highly dependent on the polyimide chemical structure and the final polymer chain orientation, as it is impacted by the anneal temperature, time and whether or not an inert or oxidizing atmosphere is present during the anneal cycle.

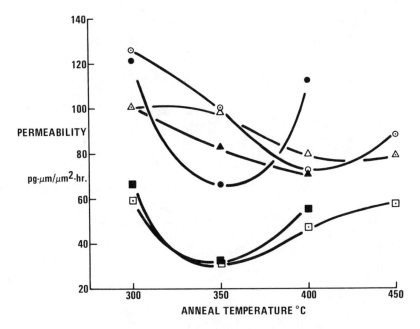

Figure 6. Moisture permeability of polyimides. Key: (polymer type) DuPont PI2545—(anneal gas) ○, N₂; ●, O₂; DuPont PI2555—△, N₂; ▲, O₂; Hitachi PIQ—□, N₂; ■, O₂.

Literature Cited

1. Ing, H.R.; Manske, R.H.F., J. Chem. Soc., 1926, 2348.
2. Frazer, A.M. "High Temperature Resistant Polymers"; John Wiley & Sons Inc.; New York, 1968.
3. Sato, K.; Harada, S.; Saiki, A.; Kimura, T.; Okubo, T.; Mukai, K., IEEE Trans. on Parts, Hybrid and Packaging 1973, 9, No. 3, 176.
4. Rubner, R.; Siemens Forsch. - u. Entwickl.-Ber., 1976, 5, 92.
5. Rothman, L.B.; J. Electrochem. Soc. 1980, 127, 2216.
6. ASTM Part 20, American Soc. for Testing Materials, 1916 Race St., Phil. Pa. 19103, 1980, Pg. 760.
7. "Pyralin Processing Bulletin PC-2"; E.I. Dupont Co., Fabrics and Finishes Dept., 1007 Market St., Wilmington, Del., 19898. Also Y.K. Lee and J.D. Craig, This volume.
8. Data to be published in J. Electrochem. Soc.
9. Gregoritsch, A.J.; Proc. IEEE Reliability Physics Sym. 1976, 14, 228.
10. Wilson, A.M.; Ext. Abs. J. Electrochem. Soc., Fall Meeting 1980, Hollywood, Fla. 1236.

RECEIVED October 19, 1981.

Implications of Electronic and Ionic Conductivities of Polyimide Films in Integrated Circuit Fabrication

GEORGE A. BROWN

Texas Instruments Inc., Dallas, TX 75265

The increasing importance of multilevel interconnection systems and surface passivation in integrated circuit fabrication has stimulated interest in polyimide films for application in silicon device processing both as multilevel insulators and overcoat layers. The ability of polyimide films to planarize stepped device geometries, as well as their thermal and chemical inertness have been previously reported, as have various physical and electrical parameters related to circuit stability and reliability in use (1, 2, 3). This paper focuses on three aspects of the electrical conductivity of polyimide (PI) films prepared from Hitachi and DuPont resins, indicating implications of each conductivity component for device reliability. The three forms of polyimide conductivity considered here are bulk electronic; ionic, associated with intentional sodium contamination; and surface or interface conductance.

Bulk Electronic Conductivity

The level of electrical conductivity of polyimide films is obviously of great importance in their potential application to integrated circuit fabrication: their primary function is as insulators between conducting leads operating at varying electrical potentials. The requirement on them goes considerably beyond the prevention of shorting or passage of gross leakage currents, however. Even very small amounts of charge transport or storage in multilevel insulator structures may cause changes in the electrical characteristics of underlying devices in the silicon surface, which could result in poor circuit performance or failure. For this reason, two aspects of the electrical conductivity are measured and correlated. The first is the normal static dc conduction measured on circular metal-polyimide-metal samples using a variable dc voltage source and an electrometer. The other approach involves measurement of the shift in threshold voltage under bias-temperature stress of insulated gate field effect transistors incorporating polyimide and silicon dioxide films in their gate insulators. The polyimide conductivity is inferred from the kinetics of the instability. The latter measurement is of special interest in that it simulates conditions potentially present in the field insulators of polyimide-multilevel circuits under reliability test conditions or in use.

DC Conduction. Cross-sectional and top views of the test structures for the dc conduction measurements are shown in Figure 1. Fabrication begins with p-type silicon wafers 3 inches in diameter, which are first doped with boron on the front

0097-6156/82/0184-0151$05.00/0

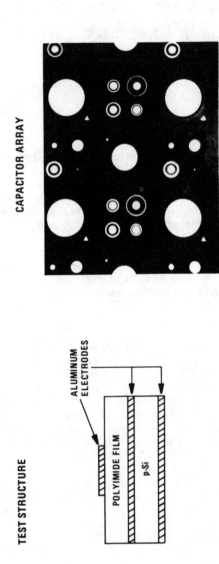

Figure 1.   Test structure array for polyimide dc conduction measurements.

and back surfaces to improve electrical contact. Aluminum-2% copper films one micrometer thick are deposited on both sides and sintered into the silicon to assure low resistance ohmic contact between the polyimide and the measuring system. The Hitachi PIQ or DuPont PI 2545 polyimide resins were spun onto the aluminized substrates to thicknesses controlled by spin speed. No adhesion promoter was required for these samples because of the good adhesion of the polyimide to the aluminum metallization. The polyimide was cured in a series of heat treatments at increasing temperatures, ending with a final conversion at 300°C for 4 hours in air. Top aluminum electrodes were evaporated from an induction heated source to minimize hot electron or x-ray radiation damage to the polyimide. The metal film was patterned as shown in Figure 1, forming a capacitor array with guarded and unguarded units varying in area from 5×10^{-5} to 2×10^{-2} cm^2.

The dc conduction measurements were made using the circuit shown in Figure 2, which illustrates the connection of the guard ring to circumvent measurement of surface leakage currents. Samples of both PIQ and DuPont PI 2545 material have displayed good ohmic I-V curves at voltages up to the destructive breakdown of the Al-PI-Al-Si sandwich structures, which occurs at field strengths up to 3 MV/cm in the polyimide. Bulk resistivities deduced from this data range up to the order of 10^{16} ohm-cm for both resins, but both also display localized regions or defects of slightly higher ohmic conductivity, from which lower values of bulk resistivity might be calculated. In addition, for certain polyimide surface preparations, high values of surface conductance have been observed. Analysis of the ring-dot structures of various radii shown in Figure 1 yields surface sheet resistance values as low as 10^6 ohms per square.

In many samples, particularly at elevated temperatures, superlinear current-voltage characteristics are seen. Most of this data seems best fitted by a log I-$V^{1/2}$ dependence, as indicated in Figure 3, which is such a plot for three PIQ samples measured at 100°C. While such a dependence may be characteristic of various charge transport mechanisms, it is reasonable to associate those in Figure 3 with Schottky emission, which is the injection of carriers over an energy barrier lowered by their image force, because of the magnitude of the slopes of the log I-$V^{1/2}$ plots. In the simplest case, assuming a uniform interface and a charge-free insulator, Schottky emission yields linear log I-$V^{1/2}$ plots with a slope of

$$\beta_S = q/kT \, (q/\pi n^2 d)^{1/2} \tag{1}$$

where
    n = index of refraction of the film
    d = film thickness
    q = electronic charge
    k = Boltzmann's constant
    T = Absolute temperature.

To test the agreement of the experimental and theoretically predicted slopes, the index of refraction inferred from equation (1) and the experimental data of Figure 3, n = 1.4, is compared with the optically determined index, n = 1.6 (4). This is reasonably good agreement in view of the material variability and simplifying assumptions involved.

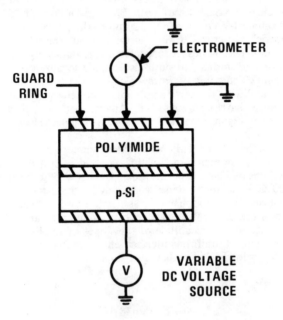

Figure 2.   The dc conduction measurement circuit.

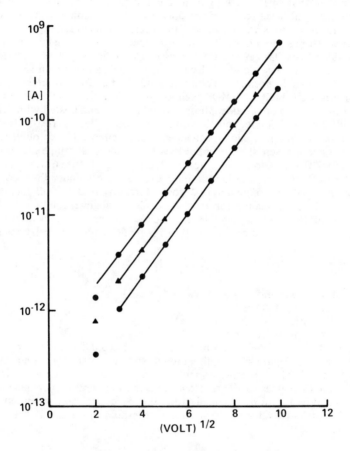

Figure 3.   The dc conduction in PIQ polyimide: log I–V$^{1/2}$ plots. Conditions: W_p = 12,870 Å; T = 100°C; and area = 0.01 cm².

To characterize the temperature dependence of this high field conductivity, the ordinate intercept current values of Schottky plots like those of Figure 3 are plotted against inverse Absolute temperature in Figure 4. The data, taken on PIQ samples at temperatures between 100°C and 200°C, is described by a thermal activation energy of about 0.7 eV.

Threshold Voltage Drift in MPOS Transistors. While the bulk resistivities quoted above are quite high, especially for organic polymers, they are nonetheless considerably lower than that of thermally grown silicon dioxide, particularly at low fields. For this reason, multilevel structures of polyimide over silicon dioxide biased under steady electric field stress will be subject to differential-conductivity-related charge transport and storage effects similar to those encountered in metal-silicon nitride-silicon dioxide-silicon devices (5). Such charge storage at the polyimide-oxide interface can potentially cause conductivity-type inversion of the silicon surface underlying second-level leads, posing a reliability hazard. To test for this effect, field oxide test transistors were built in NMOS technology with gate insulators comprised of 9000 Å of thermal silicon dioxide, phosphorus stabilized and coated with 2.2 micrometers of PI 2545 polyimide, and aluminum gate electrodes. A cross-section of this device is shown in Figure 5, along with an equivalent circuit of the double level gate insulator. These units were stressed with ± 15 volt gate biases at temperatures between 100°C and 300°C for various times up to 1200 minutes. Figure 6 shows the magnitude of the shift in field threshold voltage of the devices under these conditions. Saturating shifts of approximately 50 volts were observed for all temperatures. Assuming ohmic conductance for the polyimide, negligible conductance for the oxide, and complete polyimide-oxide interfacial charging, the field threshold shift can be derived using the equivalent circuit of Figure 5, yielding the relationship

$$\Delta V_{TFX} = - \frac{C_o}{C_p} V_s \left[1 - \exp - \left(\frac{(C_o + C_p) t}{R_p C_p C_o} \right) \right] \qquad (2)$$

where C_o, C_p = oxide and polyimide capacitances
R_p = polyimide resistance
V_s = stress voltage
t = time under stress.

The effective bulk polyimide resistivity can be extracted from comparison of the measured curves in Figure 6 with this model, and values at 100°C and 158°C are shown in Figure 7, which is a reproduction of the Arrhenius plots of Figure 4. It is seen that both the values of the resistivity and the temperature dependence are in good agreement with those obtained from the dc conduction measurements. In addition, the saturated value of the measured threshold instability is well predicted by the model.

Sodium Ion Conductivity

Because of its great natural abundance and its behavior as a mobile charge in silicon dioxide films, sodium contamination has long been a major reliability concern in silicon device processing (6). Thus, the sodium barrier properties of films considered for device fabrication are of great importance in shielding underlying oxides and pn junctions from sodium contamination associated with handling and packaging.

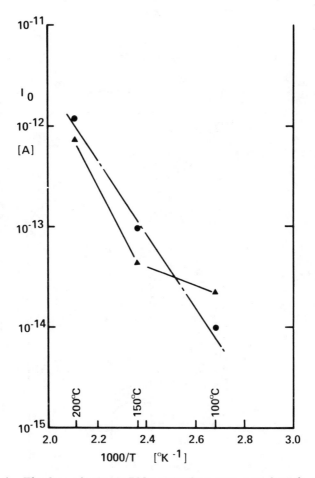

Figure 4. The dc conduction in PIQ polyimide: temperature dependence. Conditions: area = 0.01 cm²; ΔH ≅ 0.7 eV; I_o—▲, PI-2 and ●, PI-3.

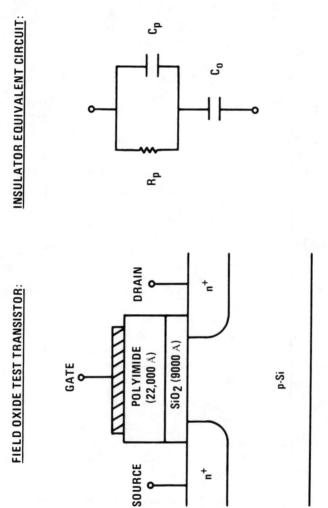

Figure 5. Polyimide–SiO₂ transistors for threshold voltage drift measurement.

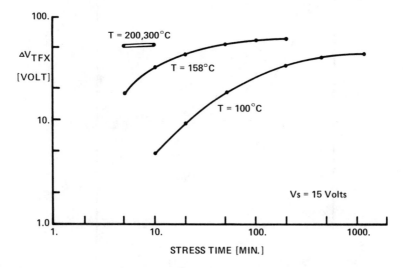

Figure 6.   Field threshold voltage instability in polyimide multilevel MOS devices.

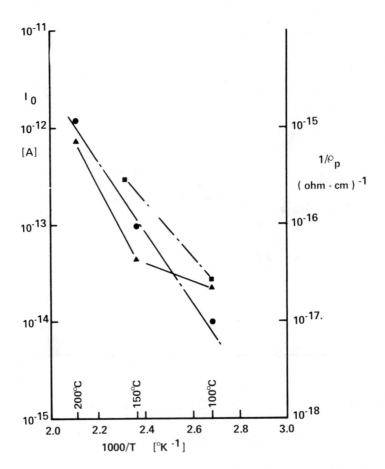

Figure 7.   Temperature dependence comparison of threshold voltage drift and dc conduction data.   Conditions: area = 0.01 cm²; ΔH ≅ 0.7 eV; I₀—▲, P I-2 and ●, P I-3; 1/ρₚ—■, ΔV_TFX data.

Sodium contamination and drift effects have traditionally been measured using static bias-temperature stress on metal-oxide-silicon (MOS) capacitors (7). This technique depends upon the perfection of the oxidized silicon interface to permit its use as a sensitive detector of charges induced in the silicon surface as a result of the density and distribution of mobile ions in the oxide above it. To measure the sodium ion barrier properties of another insulator by an analogous procedure, oxidized silicon samples would be coated with the film in question, a measured amount of sodium contamination would be placed on the surface, and a top electrode would be affixed to attempt to drift the sodium through the film with an applied dc bias voltage. Resulting inward motion of the sodium would be sensed by shifts in the MOS capacitance-voltage characteristic.

There are several difficulties in the application of this technique to the analysis of sodium barrier properties of these polyimide films. First, as we have seen above, large shifts in the surface potential characteristics of MPOS structures can be associated with electronic conduction in the polyimide and charging of the polyimide-oxide interface. These shifts are not readily separable from any that might be caused by the inward drift of sodium ions. Second, the effect of the electronic charging process is to buck out the electric field in the polyimide which is needed to drive the ion drift mechanism. As seen in Figure 6, the electric field is reduced to very small values in a matter of minutes or less, particularly at the higher temperatures where ion drift would normally be measured.

To circumvent these difficulties, a ramp voltage-integrated charge technique novel in this application has been adopted, which establishes a small but steady electric field in the polyimide for an extended period and permits separation of electronic and ionic components of conductivity. Figure 8 displays the operation of the method. If a voltage ramp is applied to a composite polyimide-SiO_2 capacitor at room temperature ($25°C$) where conductivity components are negligible, a linear charge-voltage relationship is seen having a slope equal to the series combination of the oxide and polyimide capacitances. At elevated temperatures, the slope increases, approaching the value of the oxide capacitance, which indicates that the polyimide is now acting as a resistor charging the underlying oxide capacitance. If the sample is free of mobile ionic contaminant, the Q-V plot will follow the curve A-B-C'. To measure sodium ion conductivity, a contaminant layer is applied between the polyimide surface and the metal electrode, and the sample is held under bias at point A while heating to test temperature and until the voltage across the polyimide layer is reduced to zero. When the ramp is started, a positive bias develops across the polyimide permitting ion drift toward the oxide-polyimide interface. At point B near zero volts, field reversal occurs in the oxide, and any ions collected from the polyimide or the oxide will drift rapidly toward the silicon surface as the curve goes to point C. Reversal of the ramp reverses the process. Key to the technique is the fact that field reversal in the polyimide occurs at points A and C, and in the oxide at points B and D.

Experiments like those described above have been performed to evaluate sodium ion barrier properties of Hitachi PIQ and DuPont PI 2540 polyimide films. Also included in the comparison were silicon nitride coatings plasma deposited in both tensile and compressive stress modes. The structure of the samples is illustrated in Figure 9. N-type, (111) oriented silicon substrates were cleaned and oxidized in dry oxygen ambient at $1100°C$ to form a 1060 Å SiO_2 film. Wafers intended for polyimide characterization were coated with an organic silane film (gamma glycidal amino propyl trimethoxysilane) to promote adhesion of the polyimide to the oxide surface. The polyimide resins were spun onto the wafers at speeds to produce final

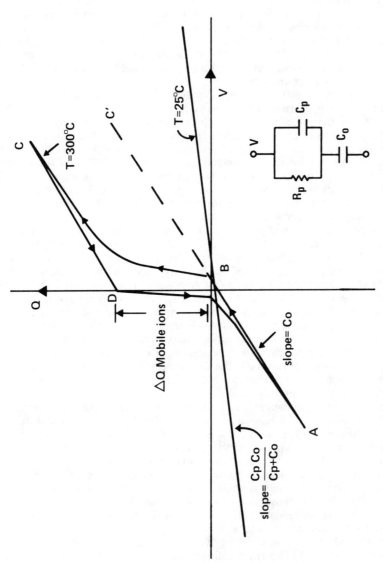

Figure 8.   Ramp voltage–integrated charge measurement technique (Q–V plot).

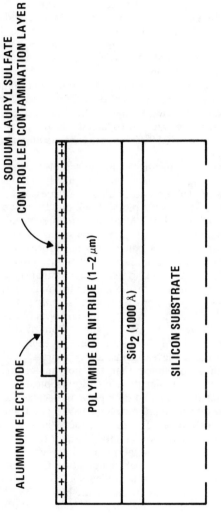

Figure 9. Test structure for sodium ion barrier measurements.

film thicknesses between one and two micrometers after curing in air at 300°C for four hours. Silicon nitride deposition was done in a plasma reactor at temperatures near 300°C, with plasma pressure and constitution varied to produce films in tensile or compressive stress relative to their substrates. Samples from each coating group along with oxide pilots were contaminated using a sodium lauryl sulfate spin-on solution calibrated for a sodium contamination level of about 2×10^{13} ions/cm^2 (8). Aluminum electrodes were applied to all wafers by evaporation from an induction heated source, and patterned as shown in Figure 1.

The ramp voltage-integrated charge measurements were made as described above, using the circuit shown in Figure 10. The voltage ramp used was a part of a Princeton Applied Research Model 410 C-V Plotter having a wide range of possible amplitudes (up to ±100 volts) and ramp rates. Most measurements were made within the ±25 volt, 0.01-1.0 volt/second amplitude and ramp rate ranges shown in the figure. The integrating electrometer used was a Keithley Model 602 electrometer operated in the "Q" mode. The test chamber, which was also used for the dc conduction and V_{TFX} drift measurements described above, is a light-tight, double ambient shielded structure with an electrically heated substrate controllable in the temperature range from 25°C to 300°C. A test temperature of 250°C was used for the sodium barrier measurements made here.

Results of the Q-V drift measurements interpreted as shown in Figure 8 are given in Table I. Each cell of the test included 3-5 wafers, and at least 3 measurements were made on each wafer. The insulator thickness values, W_{ins}, are taken from the room temperature slopes of the Q-V plots. At elevated temperatures, the slopes of all samples approximated the 1000 Å oxide thickness value in the positive and negative voltage ranges away from the charge transition. The drift measurements on the Oxide Only samples serve as a control on the process and handling contamination, and as a calibration of the sodium lauryl sulfate intentional contamination. It is believed that the 1.3×10^{11} ion/cm^2 contamination of the oxide control sample was associated primarily with the aluminum metallization process, because the nitride-coated wafers showed a greater level of stability. By comparison with this control, both polyimide films showed much higher levels of contamination, up by a factor of ten or more. However, for neither the DuPont or Hitachi material was the charge shift on the contaminated slices significantly greater than that on the uncontaminated controls, nor did the levels of instability approach those of the intentional contamination. It is thus concluded that the polyimide, once cured, provides a relatively effective ion barrier under these conditions, but that the underlying oxide may become contaminated with sodium or polar molecules during film application and curing. The silicon nitride samples also displayed good ion barrier properties in this test. They were comparatively free of the process-induced contamination found with the polyimide wafers, and showed total charge drift characteristics below the limit of resolution for the technique, about 1×10^{11} charges/cm^2 in this case. As mentioned above, the slopes of the high temperature Q-V plots for the nitride also increased to the oxide capacitance value, indicating that the nitride conductance was high in this condition. For comparison, static-bias MOS C-V stress tests were performed on some of the nitride samples to display the effect of the electronic conduction and trapping instability on samples where no evidence of ion drift was found in the Q-V plots. These results are also shown in Table II. Effective charge shifts of the order of 10^{12} charges/cm^2 were observed.

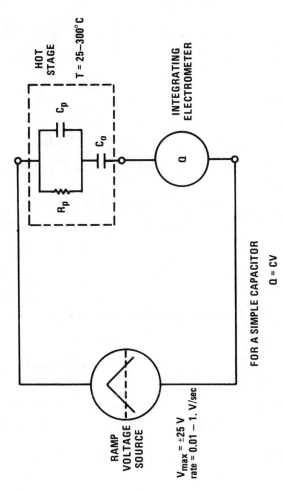

Figure 10. Ramp voltage—integrated charge measurement circuit.

Table I.   Sodium Ion Barrier Test Results for Polyimide and Silicon Nitride Films.

PROCESSING	W_{INS} (Å)	ΔQ DRIFT (IONS/CM^2)	
		CONTROL	CONTAMINATION
OXIDE ONLY	1,060	1.3E11	2.4E13
PIQ ON OXIDE	18,500	1.4E12 1.6E12 (C-V)	3E12
DUPONT 2540 ON OXIDE	13,800	8E12	7E12
COMPRESSIVE NITRIDE ON OXIDE	7,400	<1E11 8E12 (C-V)	<1E11 2.8E12 (C-V)
TENSILE NITRIDE ON OXIDE	3,000	<1E11 3E12 (C-V)	<1E11 3E12 (C-V)

ALL DRIFT MEASUREMENTS AT 250°C USING SLOW RAMP Q-V TECHNIQUE
CONTAMINATION: SODIUM LAURYL SULFATE SPIN-ON

Table II.   Lateral Interface Conductance Test Results.

OVERCOAT MATERIAL	$\Delta g_d(0)^*$ [μ mho]
POLYIMIDE (PIQ or PI2545)	$10^{-5} - 10^{-1}$
SILICON NITRIDE	$10^{-5} - 10^{-3}$
SILICON DIOXIDE (PHOS. DOPED)	$10^{-5} - 10^{-2}$
SPUTTERED QUARTZ	$1 - 10$

*STRESS CONDITIONS: $T_A = 300°C$, t = 15 MIN, $|V_s| = |V_{TFX}| + 10^V$

Lateral Interface Conductance

Lateral charge spreading on insulator surfaces and at insulator-insulator interfaces is recognized as a potential reliability problem in silicon integrated circuit operation. High levels of leakage current and even unwanted linkage of unrelated circuit nodes can result from potentials appearing between metal leads due to this effect. A standardized test has been proposed (9) as a semi-quantitative control for lateral charge spreading at field oxide-overcoat interfaces, and a relationship demonstrated between its results and those of operating life tests on integrated circuits overcoated by the same processes. This standardized test procedure makes use of the field oxide test transistor commonly found on most MOS integrated circuit chips. This test structure is shown in Figure 11, along with a plot of the initial drain conductance-gate voltage characteristics of the device as the solid curve. A detailed description of the test will not be given here, as it is available in reference (9), but the basic idea is outlined in Figure 11. After initial measurement of the g_d-V_g curve at room temperature, stress is applied by raising the potential of the gate electrode above the threshold voltage at elevated temperature giving charge an opportunity to spread at the overcoat-oxide interface as indicated by the arrows. If such charge spreading does occur, a subthreshold drain conductance characteristic like the dashed curve in the g_d-V_g plot will be apparent after stress. The zero bias drain conductance increment, Δg_{DO}, is taken as a measure of the degree of lateral charge spreading.

As a part of the evaluation of polyimide films as protective overcoat layers for integrated circuits, both NMOS and CMOS circuits have been built having test structures like those shown in Figure 11. Stress testing by the procedure described above has been carried out with the material in wafer form, using the controlled ambient test chamber described above. Stress conditions, derived in reference (9), call for a test temperature of 300°C for 15 minutes with a gate voltage 10 volts greater in magnitude than the initial field threshold voltage. A summary of the results of these tests on both DuPont PI 2545 and Hitachi PIQ polyimide overcoats is compared with ranges of values obtained in similar tests with other overcoat films; plasma deposited silicon nitride, phosphorus doped silicon dioxide, and sputtered quartz. A rather wide range of results has been obtained with the polyimide, independent of the source of the resin used. It is believed that the poorer, high conductance results are associated with contamination resulting from preparation of the field oxide surface, such as adhesion promoter application. A sufficient number of favorable results have been obtained to demonstrate that polyimide could be a competitive material for integrated circuit overcoat applications in terms of charge spreading characteristics.

Summary

Three aspects of the electrical conductivity of polyimide films applied in the fabrication of silicon integrated circuits have been considered, together with the implications of each relative to the reliability of those circuits. It may be concluded that the electronic conductivity of polyimide films may be sufficiently high to cause a serious reliability hazard due to field inversion at elevated temperatures when the polyimide is applied as a multilevel insulator in combination with silicon dioxide. Beyond the obvious approach of striving for higher resistivities in the films, this problem may be circumvented by design limitations in routing leads away from sensitive underlying device areas, limiting application of the technology to low

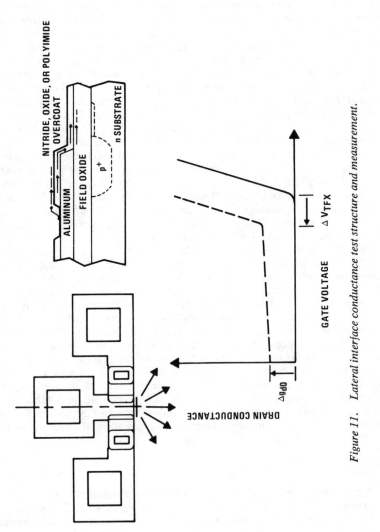

Figure 11.   Lateral interface conductance test structure and measurement.

voltage circuits or those with high surface doping densities which will reduce their sensitivity, or by employing a device fabrication procedure that incorporates some sort of field shield to isolate the surface of the active circuit from charging effects proceeding in the multilevel structure.

It appears that both the Hitachi and DuPont polyimide films, when cured, significantly impede the drift of sodium ions at normal device operating temperatures. There is, however, evidence that underlying device oxides can be contaminated by sodium and/or moisture or other polar molecules during the application and curing of the polyimide films. Cleaner resins and adequate device stabilization may control this problem. Further work will be required to characterize contamination levels associated with specific aspects of the processing, such as the adhesion promoter and the polyimide resins themselves.

The results of semi-quantitative charge spreading tests suggests that the lateral conductance of polyimide-field oxide interfaces can be sufficiently low to permit reliable device operation. This topic must be addressed in the context of the overall processing of the interface, including any adhesion promoters used.

Acknowledgements

The support of colleagues Arthur Wilson in polyimide processing and helpful discussions, and Charles Baker in preparation and application of the sodium lauryl sulfate controlled contamination solutions is gratefully acknowledged. Mary Mayfield and James Field assisted in the processing of the samples, and Ronald Huff made some of the electrical measurements. Their careful work is much appreciated.

Literature Cited

1. Sato, C.; Harada, S.; Saiki, A.; Kimura, T.; Okubo, T.; and Mukai, K. IEEE Trans. on Parts, Hybrids, and Packaging, 1973, PHP-9, 178.
2. Saiki, A.; Harada, S.; Okubo, T.; Mukai, K.; and Kimura, T. J. Electrochem. Soc. 1977, 124, 1619.
3. Gregoritsch, A. J. Reliability Physics Symposium, 14th Annual Proceedings 1976, 228.
4. Givens, L. Texas Instruments Incorporated, private communication.
5. Froman-Bentchkowsky, D.; Lenzlinger, M. J. Appl. Phys. 1969, 40, 3307.
6. Mathews, J. R.; Griffin, W. A.; Olson, K. H. J. Electrochem. Soc. 1965, 112, 899.
7. Snow, E. H.; Grove, A. S.; Deal, B. E.; Sah, C. T. J. Appl. Phys. 1965, 36, 1664.
8. Baker, C. Texas Instruments Incorporated, private communication.
9. Brown, G. A.; Lovelace, C.; Hutchins, C. Reliability Physics Symposium, 11th Annual Proceedings, 1973, 203.

RECEIVED October 19, 1981.

Improved Room-Temperature Vulcanized (RTV) Silicone Elastomers as Integrated Circuit (IC) Encapsulants

CHING-PING WONG

Western Electric Co., Inc., Engineering Research Center, Princeton, NJ 08540

Of all the commercially available organic and inorganic polymeric materials, RTV silicone elastomer has proved to be one of the most effective encapsulants used for mechanical and moisture protection of the Integrated Circuitry (IC) devices. A general overview of the RTV silicone elastomer and its commercial preparation and cure mechanism are described. Improved electrical performance of the RTV silicone encapsulant, by immobilizing the contaminant ions, such as Na^+, K^+, Cl^-, with the addition of the heterocyclic poly-ethers as the contaminant ion scavengers seems to have a potential application as the contaminant ionic migration preventor in the electronic applications.

Since World War II silicone (organosiloxane) polymers have been used in a variety of applications where properties of high thermal stability, hydrophobicity and low dielectric constant are necessary (see Figure 1). One particular application of interest to us is the use of silicones as encapsulants or conformal coatings for integrated circuits. Work in 1969 at Bell Laboratories demonstrated that silicone RTVs exhibited excellent performance as moisture protection barriers for Integrated Circuitry (IC) devices (1). Since that time, a number of different silicone RTVs have been adapted for use on ICs within the Bell System. However, as the design of ICs has steadily moved to smaller and smaller dimensions, the requirements placed on the encapsulation have risen. For example, early ICs were made with 50 to 75 micron design rules, present devices are at the 3-5 micron level and the trend is now to submicron geometries. At this level, the silicone cure mechanism, surface chemistry and level of contaminants become critical, and it is these areas that we are investigating.

0097-6156/82/0184-0171$05.00/0
© 1982 American Chemical Society

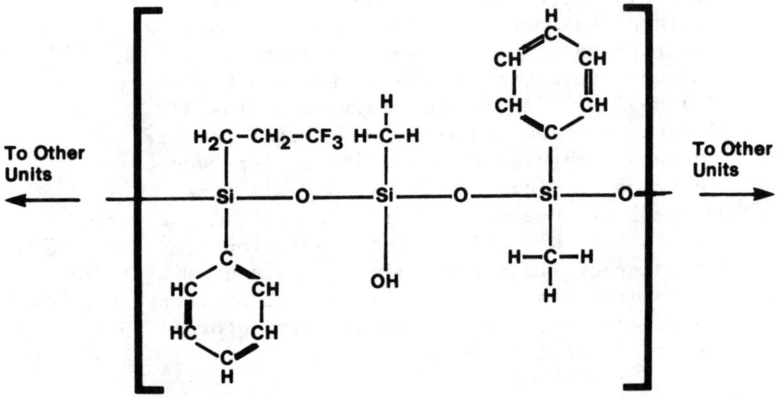

Figure 1.   Silicone structure consists of Si—O—Si backbone that provides thermal stability of the material. Hydrocarbon radicals that attach to silicon atoms provide water-repelling properties.

The primary commercial source of silicone polymers is the Rochow process wherein a stream of alkyl or aryl monohalide (typically the chloride) is passed through a heated bed of pure silicon alloyed with copper metal. The exact mechanism of this process is not well characterized but is presumed to proceed through an organocopper intermediate (2). The major product of the reaction is the diorganodihalosilane, however, some production of monoorganotrihalo- triorganohalo-tetraorgano-and tetrahalosilanes is observed. The major product is purified by distillation and is then catalytically hydrolyzed to disilanols which are unstable and combine to form a mixture of cyclic siloxane oligomers (primarily trimer and tetramer) and linear hydroxy end-blocked (HEB) siloxane polymers. The cyclic oligomers can be ring-opened and condensed into linear polymers. The molecular weight of the HEB siloxanes is controlled by the reaction conditions. End-blocking can be changed by a variety of reactions. The HEB siloxanes are typically fluids of viscosities varying from a few centistokes to resins and gums (with silicone gum the molecule weight could go up to millions). In these cases, they are not suitable as coatings and must be crosslinked or vulcanized. Two major crosslinking reaction types are available, free radical initiated and condensation curing. For the purposes of this discussion, we will limit the discussion to the condensation cures.

The condensation cures can be further divided into four sub-classes; a) the carboxylate cure, b) alkoxide cure, c) oxime cure and d) amine cure. For electronic applications, alkoxide cure is preferred. However, the alkoxide cure system has not been well defined. Figure 2 indicates one of the reasonable alkoxide cure system mechanisms. The methoxy end-blocked silicone, with the catalytic effect of some organotitanates or some other types of organotin compounds, will provide one of the methoxy end-group from the silicone polymer to react with a hydroxy group from the alumina substrate. The elimination of a methanol molecule and the formation of a silicone-oxygen-substrate bond will result. At the other end of a silicone polymer unit, another methoxy end group will react with moisture in air. This results in the elimination of an additional methanol molecule and the formation of the silicone hydroxy end-group (see Figure 2a). This silicone hydroxy end-group is reactive and further reacts with another silicone polymer's methoxy end-group results in chain propagation (see Figure 2b).

The electrical performance of the encapsulant is greatly dependent on its purity. Ionic impurities, such as sodium, potassium and chlorides, are harmful contaminants in the encapsulant. It has long been shown that ionic materials,

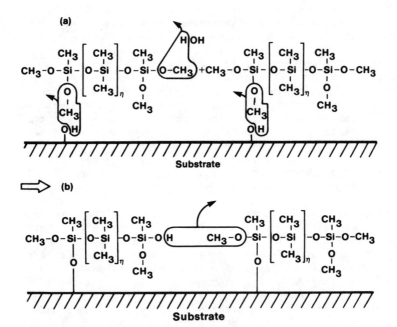

Figure 2.   Proposed alkoxide cure mechanism:silicon polymer cross-linking and bonding to surface.

(a) One of the methoxy end-groups reacts with a substrate hydroxyl group to form a Si—O substrate bond. Another methoxy end-group reacts with moisture in air to produce an active hydroxyl end-group. (b) Active hydroxyl end-group reacts intermolecularly with other methoxy group to cross-link polymers.

whether from the device surface, encapsulation materials or the environment, affect the electrical reliability of encapsulated IC devices. It has been shown (3) that exposure to hydrogen chloride accelerates the deterioration of electrical properties of silicone encapsulated triple-track conductors and resistors. Michael and Antonen report that salt atmosphere testing (M.I. Std 883A Method 1009.1) dramatically increases the Failure In Time (FIT) rate for silicone encapsulated devices over those devices which have not been exposed to salt (4). Experiments in our laboratory with silicone RTVs deliberately doped with HCl showed that FIT rates also increased. From these results, it is logical to conclude that ionic contaminants do indeed cause an increase in FIT rate, especially under hot, humid conditions.

Sodium, potassium, and chloride are the most likely ions present in materials and environment. This is mainly due to their abundance in nature. Certainly, material specifications can be made to limit the levels of these ions but this makes no provision for the reintroduction of these ions from the environment. Since the source of ions can not be eliminated, it was felt that if we could incorporate a mechanism for trapping or immobilizing these ions, the silicone RTVs would demonstrate better reliability. Our investigation of a number of different types of ion trapping compounds showed that this was the case.

In 1967, C. J. Pederson of DuPont deNemours Co. synthesized the cyclic polyethers (5). These cyclic polyethers are commonly referred to as "crown ethers" (see Figure 3). In solution, crown ethers are extremely effective ligands for a wide range of metal ions. The size of the ring cavity and the ionic radius of the metal affect the stability of the complex. Tables I and II list the cavity diameters for the crown ethers and the ionic radii of a number of metal ions (6-11).

The crown ethers and the related cryptates (cryptates were first reported by J-M Lehn (12) of France in 1969 (Figure 3)), have been used in a variety of synthetic procedures primarily because of their ability to solvate ionic materials in organic solvents (13, 14). Recently, it has been shown that crown ethers in the solid state form "sandwich" like complexes with most metals and that the counterion is also tightly bound (15). It is this evidence that suggested their use as ion traps in silicone RTV formulations.

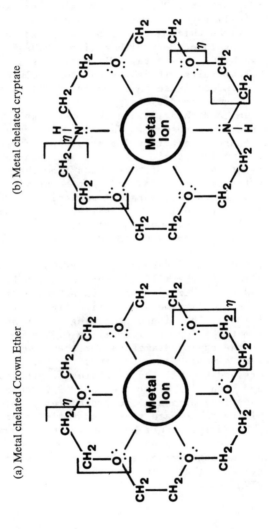

(a) Metal chelated Crown Ether

(b) Metal chelated cryptate

Figure 3.   Structure of a typical crown ether and cryptate: (a) $\eta = 1$, 18-crown-6-ether and (b) $\eta = 1$, Kryptofix 22.   Contaminant ions (such as Na^+, K^+) are immobilized (coordinated) within the cavity of the heterocycle compound.

Table I

Atomic and ionic radius of some important alkali and transition metals.

Atomic Radius (Å)	Ionic Radius (Å)		
Na 1.95	Na^+	0.95	
K 2.35	K^+	1.35	
Cu 1.28	Cu^{+1}	0.96,	Cu^{+2} 0.69
Ag 1.44	Ag^{+1}	1.26	
Au 1.46	Au^{+1}	1.37	

Table II

Structure and diameter of some crown ethers

Crown ethers	Cavity Diameter (Å)	Comments
dibenzo-12-crown-4	1.8 - 1.9	No coplanar
dibenzo-14-crown-4	1.8 - 1.9	Coplanar & symmetrical
dibenzo-15-crown-5	2.7	Coplanar & symmetrical
dibenzo-18-crown-6	4.0	Coplanar & symmetrical

Experimental Discussion

A typical encapsulation formula consists primarily of a silicone prepolymer and a crosslinking agent. Other constituents such as solvents, catalysts, fillers, and various additives may be included to enhance the application or final properties of the encapsulant. The difficulty in compounding these encapsulant materials lies in understanding the interplay of each of the components and in determining the amount and type of each one which will provide the best degree of reproducibility and reliability. Our initial experiments consisted of blending selected crosslinkers and HEB siloxanes with a commercially available silicone RTV encapsulant and monitoring the effect of the new ingredient on the electrical performance of the encapsulant. This procedure proved useful in identifying a large number of suitable components for an encapsulant formulation. This method was also used to determine effective catalysts for the alkoxide system.

After the initial screening process, it was necessary to develop model systems for both the carboxylate and alkoxide cured systems on which further experimentation could be performed. For the purposes of our testing, triacetoxymethylsilane was chosen as a crosslinker and 8000 centistoke HEB silicone as the prepolymer. Cure studies and electrical testing demonstrated that, although there is a minimum level of crosslinker to insure full cure, in excess it does not have an adverse effect. The acetoxy system is a relatively simple system to compound and, provided that the acetic acid by-product is not harmful to the metallization, it is an excellent electronic encapsulant.

On the other hand, the alkoxide system presented several problems in formulation. The system first chosen as a model consisted of a trimethoxymethyl silane crosslinker, 8000 centistoke HEB siloxane, and a catalyst. A number of catalysts were used and each exhibited different cure rates and electrical properties. DuPont tetraalkoxytitante-Tyzor[R] appears to be one of the better catalysts used in this type of curing system. Fillers are usually incorporated into the silicone formulation to improve mechanical properties, promote adhesion, and to serve as light screening and pigment agents. Cab-o-sil[R], a form of fumed silica, carbon-black, titanium dioxide and calcium carbonate are then used as RTV fillers. RTV fillers have greatly improved the RTV elastomer and have proven to be good ingredients for the IC encapsulants.

For incorporation of crown ethers and cryptates into the RTV encapsulant system as sodium and potassium ion scavengers, the total ionic contaminants must first precisely be determined. Atomic absorption is used to measure these ions in commercial silicone RTVs and silicone fluids. Values of ~10 ppm for sodium and potassium were obtained in the best samples. Chloride level was determined by potentiometric titration of the silicone with $AgNO_3$. A quantity of ion trap (either crown ethers or cryptates) was then added to the RTV silicone encapsulant, and its molar concentration was equal to the combined sodium and potassium contaminant levels.

The formulated RTV silicone is usually cured at room temperature for 16 hours and then at 120°C for 4 hours to ensure the complete removal of organic solvent. A rubbery and non-tacky elastomer is usually obtained after the curing cycle.

The electrical reliability of silicone RTV formulations was determined by using a Biased Humidity Temperature (BHT) testing procedure employing alumina ceramic IC devices with either triple track meandering resistor or conductor metallization (see Figure 4) ([3], [16]). The metallization for a resistor IC is tantalum nitride, and for a conductor, titanium-palladium-gold. These IC devices have 75 micron design parameters and, after coating with an RTV, are subjected to accelerated testing conditions. In normal testing procedures, we employ 96% relative humidity, 100°C and 180 volt dc bias to test the encapsulant. The evalution of the

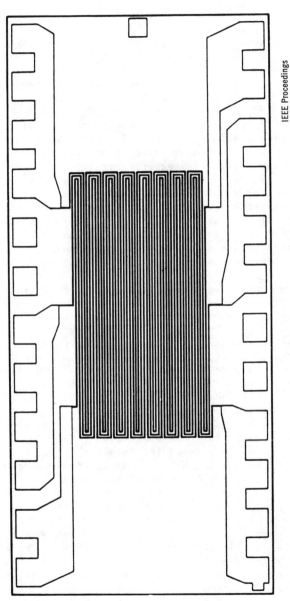

IEEE Proceedings

Figure 4.   Triple-track testing device (3).

Triple-track resistor and conductor coupons are made by deposition of Ta_2N and $Ti-Pd-Au$ metallization, respectively, on the Al_2O_3 substrate. This test pattern consists of three parallel meandering lines with 3-mil spaces between lines. Each line is approximately 3-mil wide and has 2.86×10^{-3} squares, with an overall length of 8.5 in. The number of squares of insulator between adjacent lines is approximately 3.5×10^{-4}.

encapsulant is based on their BTH testing electrical
performance. In general, the lower the resistance and leakage
current, the better the encapsulant.

Results and Discussion

To formulate a suitable RTV silicone as an IC encapsulant
based on the acetoxy cure system is relatively simple. The
acetoxy cure system typically employs an acetoxysilane as a
crosslinking agent. This reaction is rapid, even in the
absence of catalysts and produces tough, durable rubbers. The
by-product, acetic acid, can cause corrosion of aluminum
metallization on IC devices, particularly with thick coatings
where entrapment of the acid can occur. Presently available
high performance RTVs employ an alkoxide cure- the by-product
being a non-corrosive alcohol. However, this type of alkoxide
cure reaction is relatively slow even when catalyzed and
because of the need for a catalyst, the cure mechanism is more
complex than that for the acetoxy system. Nevertheless, they
all seem feasible. The results of the BHT testing on Triple
Track Resistor (TTR) testing showed that the inclusion of crown
ethers and cryptates into a silicone RTV formulation
dramatically enhances the electrical reliability of our test
vehicles (Figure 5). In the triple track resistor experiment,
we grounded the two outer tracks and biased the center track.
Then we measured the resistance change between the centers of
the conductor lines. This process measures the degree of
"electro-oxidation". Leakage currents due to impurities can
cause the resistor to anodize. The change of the resistance
with respect to the original resistance will increase with
time. This is mainly due to the oxidation process. The less
the resistant changes with testing time, the better the
encapsulant material will be. This data adds further evidence
that sodium and potassium ions contribute to the failure of
devices (17) in as much as the crown ethers with the smaller
cavities outperform those with larger cavities (18). To our
surprise, the 12-crown-4, with a cavity diameter of 1.8 Å may
be more suitable for complexing Na^+ with an ionic diameter of
1.8 Å. The 15-crown-5 at 2.7 Å may be effective for K^+ at
2.6 Å (18). In our experiment, the sodium and potassium ions
both seem to have been trapped within the crown ether quite
securely even under the most severe testing conditions.
Formation of an 'ion-pair' between the metal cation-crown ether
and halogens counterion has been observed (15). The pairing of
metal crown ether with halogen ions (i.e. cl^-) would be
beneficial in the trapping of chloride contaminant materials.
Since most halogens are potential harmful contaminants in an
encapsulant material, the formation of 'dendrites' - which

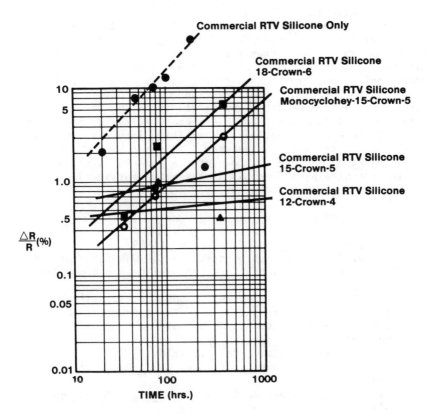

Figure 5. Triple-track resistor electrical testing performance of crown ethers in commercial RTV silicon encapsulants. Conditions: bias, 180 V; relative humidity, 96%; temperature, 100°C.

causes the leakage current between conductor path is greatly
enhanced by the presence of halogens especially under high
electrical potential bias, temperature and humidity. Such an
argument has been confirmed, and has been well documented
(3, 19).
    Cryptates have also been known to coordinate hydrochloric
acid. Hydrochloric acid and chloride ion have been associated
with the metal migration in the IC devices. The addition of
these cryptate compounds in RTV silicone encapsulant may thus
also have potential as HCl and Cl⁻ scavengers (18). This
finding may also be used to prevent silver, gold and copper
ions migrations in electronic industry applications. The
thermal and hydrolytic stability of crown ethers and cryptates
must be taken into consideration, however, when chosen as
additives. Also, since these compounds may be potentially
hazardous to our health (20), caution must be taken in using
these types of compounds. Ways to eliminate the leaching of
cryptates from the encapsulant were proposed by incorporating
the cryptate into the backbone or grafting into the substituent
side chain of the siloxane polymer. This has been shown to be
feasible (21). The use of crown ethers and cryptates to
eliminate contaminant ions may have some potential application
in electronic applications.

Literature Cited

1. White, M. L., Proc. IEEE, 57, 1610 (1969).
2. Rochow, E. G., "An Introduction to the Chemistry of the
    Silicones", 2nd Ed. New York, John Wiley and Sons, Inc.,
    1951.
3. Sbar, N., IEEE Proc, 26th Elec. Comp. Conf., 277 (1976).
4. Michael, K. W., Antonen, R. G., "The Properties of
    Silicone/Epoxy Electronic Grade Molding Compound",
    Proceedings of the Soc. of Hybrid and Microelectronics
    Conf., P. 253, Anaheim, Calif., 1978.
5. Pederson, C. J., Journal of Amer. Chem. Soc., 89, 7017
    (1967).
6. Bush, M. A., Truter, M. R., J. Chem. Soc., Chem. Comm.,
    1439 (1970).
7. Dalley, N. K., Smith, J. S., Larson, S. B., Christenson,
    J.J., Izatt, R. M., Journal of Chem. Soc., Chem. Comm.,
    43 (1975).
8. Mallison, P. R., Journal of Chem. Soc., Perkin, 261 and
    266 (1975).
9. Neman, M. A., Steiner, E. C., Van Remoirtere, F. P., Boer,
    F. D., Inorg. Chem., 14, 734 (1975).
10. Hughes, D. L., Journal of Chem. Soc., Dalton, 2374 (1975).

11.   Harman, M. E., Hart, F. A., Hursthouse, M. B., Moss, G.
      P., Raithby, P. R., Journal of Chem. Soc., Chem. Comm.,
      396 (1976).
12.   Lehn, J. M., Dietrich, B., Savvage, J. P., Tetrahedron
      Letter, 2885 (1969).
13.   Liotte, C. L., Harris, H. P., Journal of Amer. Chem. Soc.,
      96, 2250 (1974).
14.   Sam, D. J., Simmons, H. E., Journal of Amer. Chem. Soc.,
      94, 4024 (1972).
15.   Poonia, N. S., Ajaj, A. V., Chem. Rev., 29(S), 389 (1979).
16.   Mancke, R. G., The Proceedings of 31st Electronic
      Components Conference, p. 119, at Atlanta, Georgia, May
      (1978).
17.   Kaneda, A., Watanabe, Y., Japanese Patent (76-11377).
18.   Wong, C. P., "Encapsulated Electronic Devices Having
      Improved Silicone Encapsulant", U. S. Patent 4,271,425,
      June 2, 1981.
19.   DerMarderosian, A., The Proceedings of International Soc.
      for Hybrid and Microelectronics Symposium, P. 134.
      Minneapolis, Minnesota, Sept. 1978 and reference therein.
20.   Crown Ethers - PCR, Product Technical Report.
21.   Wong, C. P., "Encapsulated Electronic Devices and
      Encapsulating Compositions", U.S. Patent (allowed, in
      press).

RECEIVED October 23, 1981.

Synthesis and Properties of Branched Epoxy Resins

JON F. GEIBEL

Western Electric Co., Inc., Engineering Research Center, Princeton, NJ 08540

Epoxy resins, ubiquitous in the electronics industry, are used in a wide variety of applications in the manufacture of electronic components, including insulation materials, circuit board substrates, and component coatings and encapsulants. Thus, the cured epoxy resin is a highly functional material whose final chemical, physical, and electrical properties dictate the ultimate utility of that material.

Most coating applications mandate control over the viscosity of the epoxy resin(s) during the coating and/or curing processes. Linear epoxy resins span a large range of viscosities; however, only the upper range of commercially available resins (e.g. EEW ≥ 500 g/eq.) is suitable for powder coating applications.

An important consideration in many protective coatings is solvent resistance. It is known that decreasing the molecular weight between cross-links will decrease the solvent absorption of that thermoset (1, 2). Therefore, the lower molecular weight pre-polymers (i.e., lower EEW epoxy resins) will provide the optimum solvent resistance. These solvent resistant coatings are not accessible with higher molecular weight solid epoxy resins. To solve this problem, a higher functionality epoxy resin is typically blended with the linear epoxy resin. This class of resins is the epoxy cresol/phenol novolacs. These materials are solids, with a variety of functionalities commercially available.

Novolac resins possess low melt viscosities. This impacts on two areas: 1) When blended with a high molecular weight resin, the viscosity of the mixture is lowered. The high molecular weight resin was originally used to maintain a viscosity commensurate with processing constraints. 2) Mixing two materials with significantly different viscosities is non-trivial. Extrusion is an industrial method for blending epoxy resins. Attempting to extrude materials with widely different viscosities is difficult and often yields an inhomogeneous extrudate. A direct consequence of a poor extrusion is degraded solvent resistance.

0097-6156/82/0184-0185$05.00/0

Therefore, to eliminate the above stated properties of mixtures of high molecular weight linear epoxy resins and low molecular weight epoxy cresol novolac resins, a new method for synthesizing a class of resins possessing the wanted properties of high molecular weight and high cross-link density is presented.

Synthesis of Branched Epoxy Resins

The advancement process for synthesizing solid epoxy resins from the monomers proceeds via a fusion reaction of the diglycidyl ether of bisphenol A and bisphenol A (3). The advancement reaction is shown in Figure 1. The reaction is carried out with a calculated excess of epoxy monomer so that: 1. All the phenol is consumed; thus, the bisphenol A is the limiting reagent. 2. The product is theoretically terminated by an epoxide. Therefore, the product molecule, no matter what the molecular weight, is capable of reacting with curing agents in exactly the same manner as does the epoxy monomer.

Since both starting compounds are difunctional, this polymerization theoretically yields only linear polymers. Thus, highly functional curing agents must be used to form three-dimensional cross-linked networks. Additionally, the starting reactants are not 100% pure (i.e., they contain non-reactive end group impurities). The advancement process consumes reactive moieties and will increase the relative ratio of "dead ends" to "reactive" epoxides. This is most critical when advancing to high molecular weights, where the majority of epoxides is consumed (4).

As previously stated, to increase the cross-link density, multifunctional resins are mixed with these high molecular weight linear resins. An alternative to that approach is to add a multifunctional resin to the advancement process, thus synthesizing branched high molecular weight epoxy resins. Figure 2 outlines such a synthetic scheme.

Calculations for a branched advancement synthesis are defined to permit development of well-characterized functionality in the product molecules. In this study, the difunctional epoxy resin monomer used is the diglycidyl ether of bisphenol A, Epon 828, and the multifunctional epoxy phenol novolac resin used is DEN 438. Let:

X = Weight % Epon 828 (EEW = 190 g/eq)
Y = Weight % BPA   (PEW = 114 g/eq)
Z = Weight % DEN 438 (EEW = 183.5 g/eq)

Conservation of epoxides dictates that for a branched advancement to an EEW = 1000 g/eq. the following equation is valid:

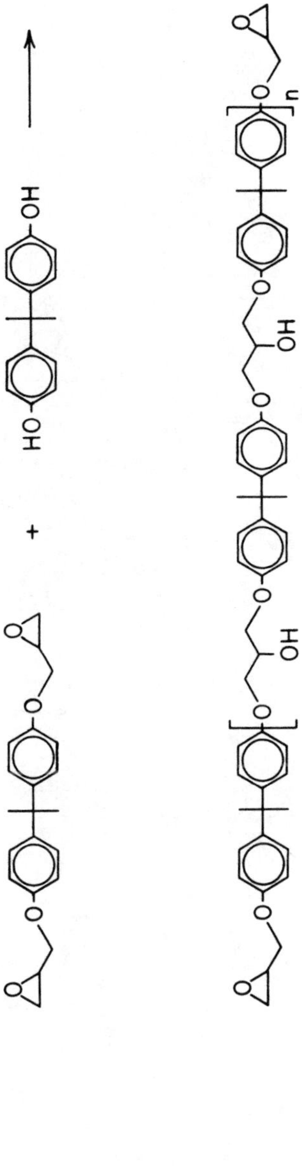

Figure 1.   Linear advancement reaction.

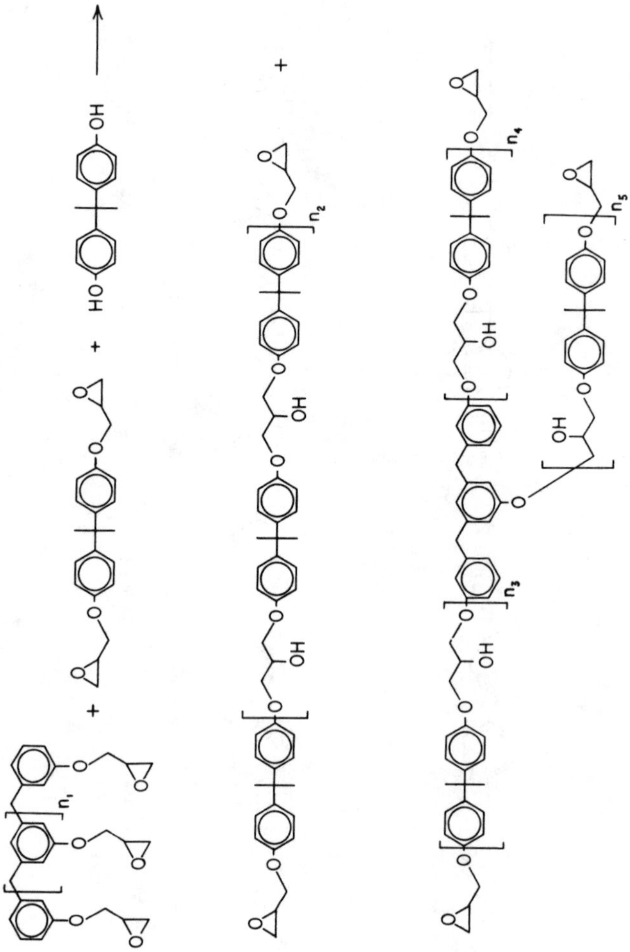

Figure 2.   Branched advancement reaction.

$$\frac{100\text{g}}{1000 \text{ g/eq}} \quad = \quad \underbrace{\frac{X\text{g}}{190 \text{ g/eq}} + \frac{Z\text{g}}{183.5 \text{ g/eq}}} \quad - \quad \frac{Y\text{g}}{114 \text{ g/eq}} \qquad (1)$$

Equivalents of epoxides remaining after advancement

Equivalents of epoxides before advancement

Equivalents of epoxides consumed by advancement

Additionally, conservation of mass dictates the sum of the weights of the reactants must equal 100%:

$$X + Y + Z = 100 \qquad (2)$$

This leaves us with 2 equations and 3 unknowns. Therefore, we define a parameter, n, such that a third equation allows a unique solution to X, Y, and Z. Thus, n is defined as the ratio of moles of branch points to moles of epoxy resin. Inspection of the first few members of an idealized homologous series shows some obvious trends.

Table I.   Idealized Homologous Series of Branched Epoxy Resins

Branch Concentration	Structure	EEW	MW
0		1000	2000
1		1000	3000
2		1000	4000

The molecular weight of the resin is clearly given by:

$$MW = 2000 + 1000n \qquad (3)$$

Thus, for 100 g of resin, the number of moles of resin is given by:

$$\text{moles of resin} = \frac{100 \text{ g}}{(2000 + 1000n) \text{ g/mole}} \qquad (4)$$

DEN 438 has a statistical functionality of 3.6 epoxides/molecule (5). Thus, incorporation of 1 molecule of DEN 438 into 1 molecule of advanced resin gives statistically 1.6 branches in that molecule of resin (3.6 epoxides are present. Two are consumed to form the linear continuation of the resin. The difference, 1.6 epoxides, is the number of branches generated.).

The titrated EEW of DEN 438 is 183.5 g/eq. Thus, the molecular weight is 3.6 times greater (660.60 g/mole of DEN 438). Thus, knowing that 1 mole of DEN 438 gives 1.6 branches, one gets a "branch equivalent weight" (BEW).

$$BEW = \frac{660.6 \text{ g}}{\text{mole of DEN 438}} \text{ X } \frac{1 \text{ mole of DEN 438}}{1.6 \text{ moles of branches}} \qquad (5)$$

$$BEW = 412.9 \text{ g/mole of branches.} \qquad (6)$$

Thus, for a given composition with Z grams of DEN 438, the total number of moles of branches is given by:

$$\text{moles of branches} = \frac{Z \text{ g}}{412.9 \text{ g/mole of branches}} \qquad (7)$$

Substituting (4) and (7) into the definition of n:

$$n = \frac{Z/412.9}{100/(2000 + 1000n)} \qquad (8)$$

Solving for Z:

$$Z = \frac{41.29 \text{ n}}{(2 + n)} \qquad (9)$$

Thus, merely by specifying n, and using equations (1), (2), and (9) a unique solution to X, Y, and Z is obtained.

Results and Discussion

An homologous series of advancement reactions was synthesized using standard reaction conditions on a 300 gram scale. The branch concentration was increased from zero to 0.677 branches/molecule. Viscosity and gel permeation (GPC) data were obtained. Table II summarizes the results.

Table II.   Branched Epoxy Resins: GPC and Melt Viscosity vs.
Branch Concentration

n	EEW	η (cp @ 175°C)	$\bar{M}_w$	$\bar{M}_n$
0	1015	2954	3242	1601
0.098	972	2918	4274	1846
0.184	971	3643	4817	1804
0.260	1030	4050	5324	1914
0.327	980	4738	5615	1911
0.441	976	5920	6499	1973
0.534	972	6449	7248	1984
0.677	1021	9839	9700	2126

The theoretical EEW was 1000 g/eq for all of these
reactions. All reactions were ± 30 g/eq (+3%) of the
calculated equivalent weight. The weight average and number
average molecular weights increase monotonically with
increasing branch concentration. The melt viscosities also
increase with increasing branch concentration.

Figure 3 shows the melt viscosity at 175°C versus branch
concentration. The melt viscosity increases by greater than a
factor of three for this homologous series of resins. At
branch concentrations greater than 0.5 branches/molecule there
is a deviation from the linearly increasing viscosity seen at
lower branch concentrations. Figure 4 displays $\bar{M}_w$ versus
branch concentration. At branch concentrations greater than
0.5, $\bar{M}_w$ increases more rapidly than at lower branch
concentrations. Figure 5 plots log (melt viscosity) versus
log($\bar{M}_w$). The linearity of this plot indicate that the size
of these epoxy resins is below the critical molecular weight
for chain-chain entanglements. The slope of this plot is 1.4
and is consistent with no chain entanglements.

An homologous series of cured coatings of branched epoxy
resins was tested for the kinetics of swelling with methyl
ethyl ketone. The series of epoxy resins was synthesized with
constant EEW and increasing branch content (EEW = 1000 g/eq and
n = 0, 0.25 and 0.50, respectively). The syntheses were
carried out on a 3.5 kg scale. Samples were cured with a
stoichiometric amount of methylene dianiline at 200°C for 30
minutes. Increasing branch content decreases the rate of
solvent absorption dramatically. The increased number of
cross-links in the cured samples of the highly branched epoxy
resins results in decreased chain mobility which is responsible
for the decreased solvent uptake. Figure 6 shows the weight
uptake of methyl ethyl ketone per unit area versus square root
of time of immersion in the neat solvent.

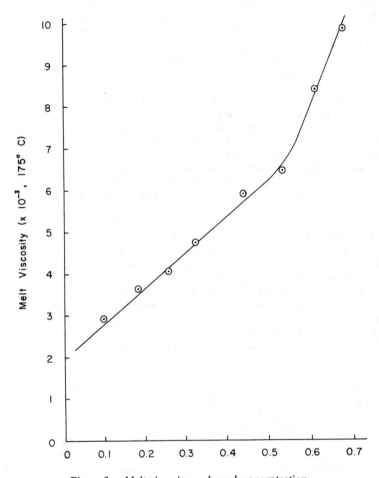

Figure 3.   Melt viscosity vs. branch concentration.

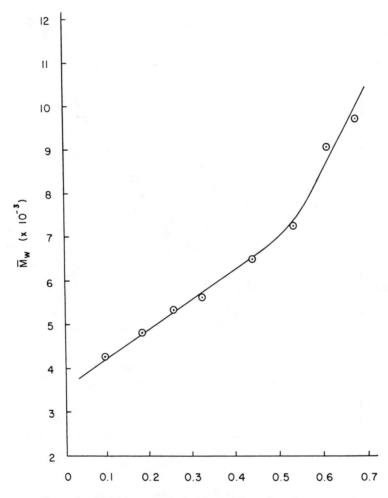

Figure 4. Weight-average molecular weight vs. branch concentration.

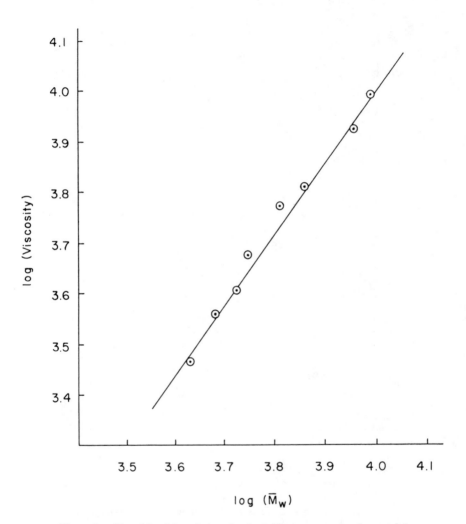

Figure 5.   Plot of log (viscosity) vs. log (weight-average molecular weight).

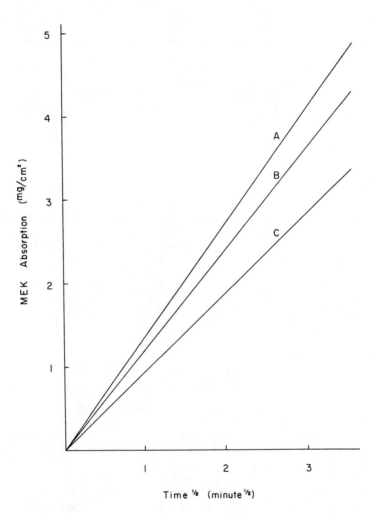

Figure 6. MEK absorption vs. square root of time. Key: A, n = 0.00; B, n = 0.25; and C, n = 0.50.

Conclusions

A concise synthesis of branched epoxy resins is presented. Stoichiometric calculations are discussed which treat the synthesis in an idealized statistical model. The calculations can be adapted to any well characterized reactants.

An homologous series of epoxy resins with constant epoxide equivalent weight and increasing branch concentration displayed increased melt viscosity and weight average molecular weight. The range of molecular weights investigated in this study was shown to be below that required for chain-chain entanglements.

Powder coatings were prepared by extruding stoichiometric quantities of methylene dianiline with epoxy resins varying only in branch concentration. More highly branched resins result in a more tightly cross-linked thermoset network. Solvent absorption data indicate decreased absorption with increasing branch concentration.

Experimental

Advancement reactions were carried out on a 300 gram scale, unless otherwise specified, using commercially supplied Epon 828, BPA and DEN 438. All resins were used as received, with no further purification. Triphenylphosphine was used as received from Polysciences (lot 2-2401).

Advancement reactions were performed using electric heating mantles. Temperatures of the resin, mantle/beaker interface, and mantle interior were recorded via thermocouples. Reaction mixtures were not protected from atmospheric oxygen. The advancement reaction was initiated at 120°C using 0.1% (weight/weight) triphenylphosphine. Reaction mixtures exothermed to 175° - 185°C and were allowed to cool to 175°C. All reactions were thermally quenched by pouring out on aluminum foil exactly 30 minutes after the peak of the exotherm.

The direct titration of the epoxide equivalent weight (EEW) was accomplished by the method of Jay (7). Melt viscosity measurements were determined at 175°C using a Haake RV-3 Rotoviscometer equipped with high temperature range cone-plate viscosity sensor system (PK-401W). Gel permeation chromatography was performed on a Waters Associates HPLC using μ-styragel columns (10^5, 10^4, 10^3, 500, 500, 100, 100 Å) in series. The flow rate was 2.0 ml/min. Samples were dissolved in a spectral grade THF and filtered prior to chromatography through Millipore 0.5 μ filters (FHLP 01300). All calculations were based on the output of a differential refractomer detector.

Powder coatings of branched epoxy resins were prepared by extruding stoichiometric amounts of the epoxy resin and methylene dianiline (1 equivalent of epoxide per 1 N-H equivalent) in a Buss PR-46 single screw extruder. The extrudate was pulverized in a Bantam Micropul hammermill. Samples for solvent absorption (2) were prepared by electrostatically coating copper foil and curing for the requisite schedule. The sample weight is determined accurately prior to testing. The sample is submerged in neat MEK in a constant temperature bath at 23+0.5°C. The sample is removed after the appropriate immersion time, blotted dry, and weighed precisely 45 seconds after removal from the solvent. The absorption is calculated in $^{mg}/cm^2$ according to the following formula:

$$\text{Absorption } (^{mg}/cm^2) = \frac{Wt_2 - Wt_1}{\text{Area}} \qquad (10)$$

where Wt_2 is the weight after absorption, Wt_1 is the weight prior to absorption, and Area is the surface area of the sample. The absorption values reported are the average of four samples.

Literature Cited

1. Nielsen, L. E. Macromol. Sci. - Revs. Macromol. Chem. 1969, C3(1), 69.
2. Romanchick, W. A., Geibel, Jon F., Second Chemical Congress of the North American Continent, Las Vegas, 1980, ORPL-118.
3. Batzer, H.; Zahir, S. A. J. Appl. Poly. Sci. 1975, 19, 585.
4. Ravindranath, K.; Ghandi, K. S. J. Appl. Poly. Sci. 1979, 24, 1115.
5. Dow Chemical Company, Production Description - DEN Epoxy Novolac Resins, form No. 170 - 143B (1967).
6. Nielsen, L. E. "Mechanical Properties of Polymers", Reinhold, London, 1962, p. 58.
7. Jay, R. R., Analytical Chemistry 1964, 35, 197.

RECEIVED November 6, 1981.

Characterization of Cured Epoxy Powder Coatings by Solvent Absorption

W. A. ROMANCHICK and J. F. GEIBEL

Western Electric Co., Inc., Engineering Research Center, Princeton, NJ 08540

Crosslinks are extremely important in determining the physical properties of thermoset polymers because they increase the molecular weight as well as limit the motion of chains with respect to one another.(1) Although insoluble, a polymeric network will usually absorb and be swelled by solvents in which the uncrosslinked (uncured) polymer is soluble. The tendency to absorb solvents decreases as the degree of crosslinking is increased. This paper will describe a standard procedure for measuring MEK (Methylethyl Ketone) absorption of cured epoxy powder coatings. The test has proven to be a sensitive measure of both material properties and the effectiveness of manufacture of powder coatings. The effects of time and temperature of MEK, homogeneity of extrusion, state of cure, Epoxy Equivalent Weight (EEW), and CTBN (Carboxy-Terminated Butadiene-Acrylonitrile Copolymer) elastomer concentration on MEK absorption are discussed.

DISCUSSION OF TEST PARAMETERS

Although the detailed procedure for the MEK absorption test is given in the experimental section, certain parameters require additional elaboration. In particular, the temperature of the MEK must be tightly controlled and the coating thickness must be kept above a minimum value if solvent absorption data are to be reproducible.

As shown in Table I, the MEK absorption changes drastically for small changes in temperature. As a result, standard test samples are run in a constant temperature bath at 23°C.

0097-6156/82/0184-0199$05.00/0

Table I. MEK Absorption versus Temperature

Temperature of MEK (°C)	MEK Absorption (mg/cm^2, 5 min. dip)
18	2.996
23	3.866
28	4.298
35	4.862

Note: Linear epoxy resin, EEW = 1185 g/eq., 10% CTBN
rubber, cured with P-108 at 200°C for 15 minutes.

The second critical parameter which must be controlled is
coating thickness. The coating must be sufficiently thick on
test samples to preclude the possibility of the solvent
diffusing through to the polymer-metal interface. In a given
EEW range (600-2000 g/eq.), 6 mils of coating has proven to
be sufficient to prevent saturation. A standard powder coating
(EEW = 1173 g/eq.) was coated at various thicknesses and
cured with P-108 at 200°C for 15 minutes. The MEK absorption
for a 5 minute immersion was then measured. The results are
shown in Table II and Figure 1.

Table II. MEK Absorption versus Coating Thickness

Coating Thickness (mils)	MEK Absorption (mg/cm^2)
0.4	0.24
1.0	0.91
2.0	1.93
4.0	4.35
5.0	4.80
6.0	4.90
6.2	5.00
10.0	5.10
11.5	5.10
17.5	5.28
20.0	5.01

Clearly, the MEK absorption levels out at coating thicknesses
greater than 6 mils. Sample coatings less than 6 mils lead to
erroneous results.

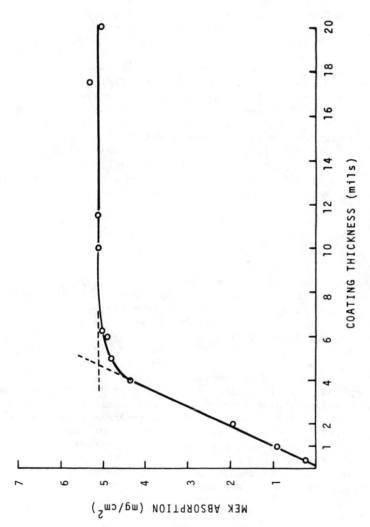

Figure 1.   MEK absorption vs. coating thickness.

MEK ABSORPTION VERSUS TIME

The absorption of MEK by cured epoxy resins follows a square
root of time dependence. Figure 2 shows a typical plot of MEK
absorbed versus square root of time. This epoxy resin (EEW = 1060
g/eq.) was cured with P-108 at 200°C for 15 minutes. Each of
the data points represents one separate experiment. In this case,
four samples were immersed in MEK, and samples withdrawn at 1, 3,
5, and 10 minutes. Each sample was weighed as described in the
experimental section. The line drawn through the data does not
pass through the origin as one would expect. This is probably a
result of waiting 45 seconds from the time the sample is withdrawn
from the MEK to the time the weight is recorded. The weight of the
swollen samples constantly decreases owing to evaporative loss of
MEK. Thus, we systematize this evaporative error by weighing the
sample exactly 45 seconds after withdrawing it from the solvent.
Presumably the swollen surface is similar for all times subsequent
to the time that the surface attains equilibrium swelling.
Therefore, the evaporative loss should be similar for the 1, 3, 5,
and 10 minute samples, simply displacing the plot downward. This
results in a negative intercept shown in Figure 2.

MEK ABSORPTION AS A MEASURE OF HOMOGENEITY

In the evaluation of functional powder coatings, it must be
noted that the final material properties depend not only on the
formulation but on the homogeneity of the material. Optimum
dispersion of resin and curing agent is necessary to form a
uniformly crosslinked network. The most widely used industrial
method for mixing solid resin and curing agent is extrusion.
Solvent absorption has proven to be a sensitive tool in evaluating
the effectiveness of extrusion (homogeneity). Table III shows the
MEK absorption of twelve different powder formulations after one,
two, and sometimes three extrusions. As can easily by seen in all
cases solvent absorption is reduced by the second extrusion while
the third extrusion does not seem to lower it further. Thus, MEK
absorption can be used as a test to optimize extruder conditions.
If the solvent absorption can be minimized for a particular
formulation in one extrusion, quality is preserved while the
manufacturing sequence is optimized.

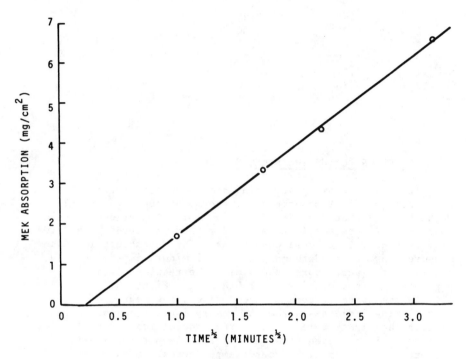

Figure 2.   MEK absorption vs. time$^{1/2}$.

Table III. MEK Absorption versus the Number of Extrusions

MEK Absorption (mg/cm^2)

Powder	1 Extrusion	2 Extrusions	3 Extrusions
1	4.44	4.09	4.08
2	5.14	4.83	
3	3.86	3.66	
4	5.10	4.47	
5	2.94	2.58	
6	1.72	1.65	
7	4.42	3.95	
8	4.30	4.00	
9	4.00	3.55	
10	0.58	0.26	
11	3.21	3.05	
12	5.30	4.70	

MEK ABSORPTION VERSUS THE STATE OF CURE

It is well known that curing times and temperatures influence the final properties of thermoset epoxy resins.(2) We have found that the MEK absorption for a linear epoxy resin containing 12% carboxyl- terminated rubber and cured with P-108 is very sensitive to cure conditions. The MEK absorption was measured in the standard manner for a 5 minute immersion as a function of several cure schedules. The results of these experiments are summarized in Table IV and Figure 3.

As the state of cure advances (i.e., the degree of crosslinking increases), the MEK absorption decreases. Note that at the higher cure temperatures, the ultimate MEK absorption is achieved more rapidly. It is not necessarily advantageous to use higher cure temperatures, as this can lead to side reactions (e.g., oxidative degradation, decomposition of dicy, changes in rubber particle morphology) which can also influence the final cured properties. It is, however, clear that the extent of cure can be easily followed by solvent absorption measurements. A necessary corollary is that comparison of the MEK absorptions of two or more formulations must be done at constant curing conditions. To compare solvent absorptions of two formulations cured at different conditions is meaningless because any differences in solvent absorptions cannot be definitively ascribed to either the cure schedule or the composition of the formulation. Thus, to be a formulating aid, solvent absorption studies mandate a high degree of consistency in the generation of samples.

Table IV.   MEK Absorption versus Cure Schedule

170°C Cure Cure Time(min)	MEK Abs(mg/cm^2)	200°C Cure Cure Time(min)	MEK Abs(mg/cm^2)	230°C Cure Cure Time(min)	MEK Abs(mg/cm^2)
5	6.95	5	5.10	3	5.35
10	5.60	10	5.00	5	5.30
30	4.80	15	4.95	15	4.70
60	4.65	30	4.65	30	4.10
90	4.50	60	4.40		
120	5.00				
180	4.05				
240	4.10				

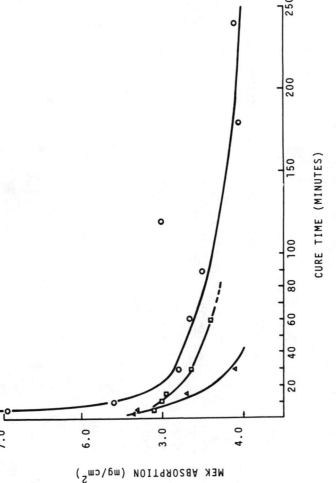

Figure 3. MEK absorption vs. cure schedule. Key: ○, *170°C cure;* □, *200°C cure; and* △, *230°C cure.*

MEK ABSORPTION VERSUS EEW OF EPOXY RESIN

Several epoxy powders were formulated in a homologous series where the EEW of the base resin changed from 858 g/eq. to 1487 g/eq. All formulations contained 10% carboxyl-terminated rubber, were extruded two times, and were cured with calculated amounts of P-108. The standard cure schedule of 200°C for 15 minutes was used. MEK absorptions were measured in the usual manner. A definite dependence of MEK absorption on EEW was found. Table V summarizes the data.

Table V.   MEK Absorption versus EEW of Base Resin

EEW	MEK Absorption (5 min)	MEK Absorption (10 min)
858	3.0	4.9
1081	3.21	5.11
1099	3.13	5.01
1216	3.90	6.00
1487	4.39	6.67

These data are consistent with the observation (3) that increased crosslinking (i.e., epoxy resins with lower EEW's) will result in a decreased propensity for the cured polymer to absorb solvents. Thus, the solvent resistence of an organic coating can be controlled by the formulator via variations in the molecular weight between crosslinks.

MEK ABSORPTION VERSUS CTBN RUBBER CONCENTRATION

The solvent swelling behavior of cured CTBN-modified epoxy resins is dramatically different than the analogous cured non-modified epoxy resins. A homologous series of formulations containing various types and concentrations of CTBN elastomers was designed to elucidate the relationship between the composition of the formulation and the solvent swelling properties of the cured resin. All formulations were extruded two times to insure homogeneity, and cured with P-108 at 200°C for 15 minutes. MEK absorption was measured in the usual manner. All of the EEW's of the base resin were calculated to be 1000 g/eq., however, variations in each resin synthesis resulted in some scatter in the EEW's. Table VI and Figure 4 summarize the MEK absorption data for the various rubber-modified epoxy thermosets.

The MEK absorption of these cured resins is a strong function of the rubber content. Increasing the weight percent of rubber increases the amount of MEK absorbed very dramatically. Curves 1 and 2 are data for 5 minute immersions, while curves 3 and 4 are for 10 minute immersions. The

Table VI.   MEK Absorption versus Rubber Type and Concentration

Rubber	Concentration(w/w %)	EEW of Base Resin(g/eq)	MEK Absorption(mg/cm^2)	
			5 min immersion	10 min immersion
None	- - -	945	1.70	2.90
X13	2.5	980	2.35	3.73
X13	5.0	1043	3.05	4.98
X13	10.0	1081	3.21	5.11
X8	5.0	969	3.03	4.67
X8	10.0	991	3.82	5.78

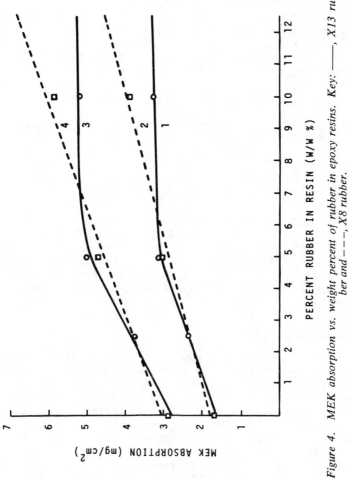

Figure 4. MEK absorption vs. weight percent of rubber in epoxy resins. Key: ———, X13 rubber and – – –, X8 rubber.

homologous series of X8 containing powders displays a monotonic
upward trend. The X13 containing powders are characterized by
a similar trend up to 5% rubber. The similarity in solvent
absorption of the 5% and 10% X13 powders is striking and may be
related to the degree of phase-separation or domain morphology.

An additional effect must be noted: As the weight percent
of rubber increases, the EEW of the resin also increases. This
was shown in a previous section to impact on the MEK
absorption. Within this rubber series the maximum difference
in EEW's is 140 g/eq. The effect of changing the EEW by 140
g/eq. on the solvent absorption is small compared to the
changes observed.

Table VII summarizes the solubility parameters of MEK, the
CTBN elastomers, and an epoxy resin of similar EEW.

Table VII. Solubility Parameters

Material	Solubility Parameter (cal/cm^3)	Reference
MEK	9.04	(4)
CTBN X8	8.77	(5)
CTBN X13	9.14	(5)
Epon 1004	8.5 - 13.3 (Average = 10.9)	(6)

It is clear, that MEK is a "good" solvent for both the
elastomers and the epoxy resin. Note that at 10% rubber, the
MEK absorption nearly doubles. This implies that a much higher
concentration of MEK is present in the rubber phase than in the
epoxy phase. This is possible because the MEK diffuses more
rapidly into the rubbery CTBN phase owing to its greater
segmental thermal motion.

CONCLUSIONS

Solvent absorption measurement has been shown to be a
sensitive and useful test method in the manufacture of epoxy
powder coatings. A test method was defined and the effects of
time and temperature of immersion described. It was shown that
solvent absorption is a measure of raw material properties (EEW
of the epoxy resin, and CTBN elastomer type and concentration),
the homogeneity of the extrudate, as well as the state of
cure. The information obtained from solvent absorption
measurements has proven to be extremely important not only in
quality control analysis but also in providing an insight into
the structure function relationships in epoxy resin chemistry.

EXPERIMENTAL

Preparation of sample - A 6" x 8" piece of 1 oz. rolled
copper is cut and folded to give a 6" x 4" sample. The open

end and sides are clipped together to prevent powder from coating the interior copper surface. The sample is then electrostatically coated with the powder to be tested. It is then cured at 200°C for fifteen minutes in a convection oven and measured with a micrometer caliper to assure a minimum coating thickness of 6 mils (0.006 inch). The clips are removed and the sample unfolded. A 2 1/2" x 2 1/2" square is cut from the center of each face. This is done to minimize coating thickness variations and edge effects.

Absorption test - The sample weight is determined accurately prior to testing. The sample is submerged in 100% MEK in a constant temperature bath at 23 ± 0.5°C. Standard immersion times are 5 and 10 minutes. The sample is removed, blotted to remove surface liquid and weighed precisely 45 seconds after removal from the solvent. The absorption is calculated in mg/cm^2 according to the following formula:

$$\text{Absorption (mg/cm}^2) = \frac{wt_2 - wt_1}{\text{Area}}$$

where wt_2 is the weight after absorption, wt_1 the initial sample weight and the area in this case, 40.3225 cm^2. The values shown in this paper are the average of four samples and are reported as a function of surface area (mg/cm^2) in contrast to a weight percent increase. Calculations based on weight percent gain suffer from imprecision due to inherent variations in sample weight and geometry. Data based on solvent uptake per unit area (exposed to the solvent) minimize these variations.

GLOSSARY

CTBN: Carboxyl-Terminated Butadiene Acrylonitrile Rubber

EEW: Epoxide equivalent weight.

MEK: Methyl ethyl ketone or 2-butanone

P-108:   Proprietary curing agent of Shell Chemical, consisting of an imidazole "accelerated" dicyanodiamide (7).

LITERATURE CITED

1.   Graessley, W. W., Accts. Chem. Res., 1977, 10, 332.
2.   Manzione, L. T., Ph.D. Dissertation, Princeton University, 1979.
3.   Nielsen, L. E., Macromol. Sci.-Revs. Macromol. Chem., 1969, C3(1), 69.

4. Billmeyer, F. W., Jr., "Textbook of Polymer Science"; John
    Wiley and Sons, Inc., New York, 1971; p.25.
5. B. F. Goodrich Co., Product Description RLP-1 (Hycar
    Reactive Liquid Polymers), Cleveland, Ohio.
6. Bekales, N. M., Mark, H. F., Gaylord, N. G., Eds.;
    "Encyclopedia of Polymer Science and Technology"; John
    Wiley and Sons, Inc., New York, 1970; Vol. 12, p. 618-626.
7. Carey, J. E., Private Communication, March, 1978.

RECEIVED October 19, 1981.

Thermal Degradation of Polymers for Molded Integrated Circuit (IC) Devices

The Effect of a Flame Retardant

R. M. LUM and L. G. FEINSTEIN[1]

Bell Laboratories, Allentown, PA 18103

This paper reports the results of a molecular-level investigation of the effects of flame retardant additives on the thermal dedomposition of thermoset molding compounds used for encapsulation of IC devices, and their implications to the reliability of devices in molded plastic packages. In particular, semiconductor grade novolac epoxy and silicone-epoxy based resins and an electrical grade novolac epoxy formulation are compared. This work is an extension of a previous study[1] of an epoxy encapsulant to flame retarded and non-flame retarded sample pairs of novolac epoxy and silicone-epoxy compounds. The results of this work are correlated with separate studies on device aging[2,3], where appropriate.

Materials

Three classes of polymer encapsulant materials were studied. These are listed in Table I and include novolac epoxy and silicone-epoxy compounds. A pure silicone formulation served as a control for comparison of the thermal degradion properties

TABLE I
Sample Pairs Compared in this Study

A. Electrical/Electronic Grade Novolac Epoxy

Sample A:	commercial composition	-FR
Sample B:	experimental lot	-1/2 FR
Sample C:	experimental lot	-non FR

B. Semiconductor Grade Silicone-Epoxy

Sample D:	commercial comppsition	-FR
Sample E:	identical to above	-non FR

C. Semiconductor Grade Novolac Epoxies

Sample F:	commercial composition F	-FR
Sample G:	commercial composition G	-non FR

(except for FR, compositions G and F are similar)

[1] Current address: INMOS Co., Colorado Springs, CO.

0097-6156/82/0184-0213$05.00/0

of the silicone-epoxy compounds. To improve their mechanical
and thermal properties, the epoxy and silicne-epoxy molding
compounds contain up to 75% by weight of a silica filler, plus
various sabilizer and processing additives. Investigations
of the epoxy and silicone-epoxy materials are reported for
compound formulations both with and without the flame retardant,
a tetrabromobisphenol-A and Sb_2O_3 combination.

   Novolac epoxy resins are produced by reaction of novolac,
a phenolformaldehyde resin, with epichlorohyrin and a base[4].
A detailed discussion of the chemistry, production and physical
properties of epoxy resins in terms of their application to
molding compounds has recently been given by Helfand and Villani[5].

   The silcone resin is a poly(methylphenylsiloxane) synthesized from methylchlorosilanes and phenylsilanes[6]. Incorporation
of phenyl groups improves the thermal stability of the silicone
resin. Epoxidation of the resin is most likely accomplished
through substitution of the oxirane ring at variou phenyl
group sites, yeilding an epoxide/siloxane ratio near unity.

   The epoxy and silicone-epoxy resins were molded into 37-mm
diameter by 4-mm thick discs. These were ground in a SPEX
model 8004 carbide lined two-ball grinder, and the resulting
powder sieved to provide controlled sample configurations for
laboratory analyses.

Apparatus

   Evolved gas analysis (EGA). Temperature programmed
(5°C/min) mass spectrometric (MS) techniques[1,7] were used to
analyze the volatile products formed during sample heating.
The polymer sample (35 mg) was pyrolyzed in a quartz cell which
was directly attached to the inlet flange of a quadrupole mass
spectrometer. Gases evolved from the polymer compound were
dynamically sampled via a 1.0-mm diameter orifice, formed into
a modulated molecular beam, and mass analyzed. Information was
obtained on the total yield of volatile products, product
composition, and individual product yields as a function of
temperature.

   Differential Scanning Calorimetry (DSC). A DuPont 990
thermal analyzer equipped with a DSC cell was employed to
record the endothermic and exothermic reactions which occurred
during temperature-programmed (10°C/min) heating of the polymer
samples. Sample weights were 15 mg, and the ambient atmosphere
was either prepurified nitrogen or line air.

   Elemental Analysis. A Phillips PW 1410/70 X-ray fluorescence spectrometer with Cr radiation was used to measure the
relative quantities of Br in the molded polymer samples.
Extractable bromide and chloride ions were detected with specific
ion electrodes after a 48-hour, 120°C steam bomb extraction.

Electrical/Electronic Grade Novolac Epoxy Results

TGA. Weight-loss measurements for the electrical-electronic grade novolac epoxy were reported in our earlier work[1]. For samples heated in N_2 to 350°C, no differences attributable to the presence of FR were observed. Isothermal measurements indicated a 20% weight loss for the unfilled molding compound after 12-days at 220°C.

EGA. A detailed discussion of the evolved gas measurements for the electrical grade epoxy compound was given in our earlier work[1]. Several sources were found to contribute to the observed outgassing: decomposition of the epoxy polymer structure; re release of trapped impurity or residual species from the poly- merization process; and breakdown of the brominated flame retardant. Evolution of volatile species from each source, however, occurred in different temperature regimes. Aromatic species and chloride products remaining from polymer synthesis dominated the initial outgassing (100-225°C). The major release of bromine products form the flame retardant did not occur until higher temperatures (275-350°C). This is well above the range employed in device testing.
The EGA data indicated that the flame retardant had little effect on the composition or formation kinetics of the volatile decomposition products. The major species released from the FR was HBr. A small amount of CH_3Br was also detected. Individual profiles of the ion signals characteristic of the HBr and CH_3Br flame retardant products are shown in Figure 1 for sample A (solid curves), and compared with the corresponding ion signal intensities observed for sample C (dashed curves). Release of HBr from sample A is shown in Figure 1c, superimposed on an increasing background signal. The temperature of maximum HBr evolution, 325°C, is in good agreement with that previously observed in this laboratory for polycarbonate samples containing tetrabromobisphenol-A, and with recent weight loss measurements[8] on the flame retardant compound itself. The observed background at the component signals used to characterize HBr (m/e=79, 80, 81 and 82) arises from fragmentation of aromatic species produced during chain scission of the epoxy polymer backbone.
Ion signals for the $CH_3{}^{79}Br^+$ (m/e=94) and $CH_3{}^{81}Br^+$ (m/e=96) isotopic components are presented separately in Figures 1a and 1b, because of the interference of the m/e=94 phenol signal. Barely detectable signal levels are observed for these ions below 300° from sample C, while the profiles from sample A provide clear evidence for the release of methyl bromide from the flame retardant. This release occurs in two stages, with the signal intensities in the high temperature stage (290°C) approximately twice those observed at the low temperature peak (190°). However, at 190°C methyl chloride is the dominant volatile component, as shown in Figure 1d, exceeding evolution

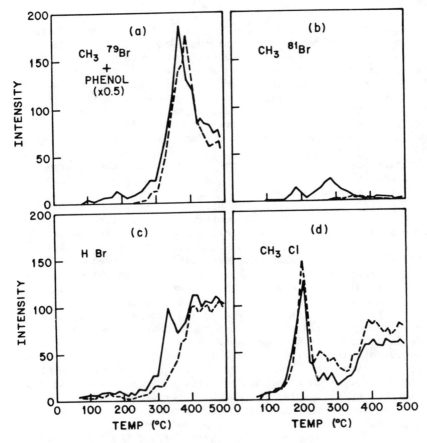

Figure 1.   *Ion profiles representative of HBr and CH₃Br flame retardant species.*
Key: ———, Sample A(FR) and – – –, Sample C(no FR).

of methyl bromide from the flame retardant by a factor of approximately ten.

DSC. Flame retardant effects on the decomposition chemistry of molding compounds can also be detected from DSC measurements of the heat released during exothermic reactions of constituents of the polymer blend. The DSC data characteristic of the electrical grade epoxy samples has been discussed in detail previously[1], and is reproduced in Figure 2 for comparison with the other sample pairs. The effect of the flame retardant is evidenced by the exotherm at approximately 350°C which increases with increasing FR content.

Elemental Analyses. X-ray fluorescence measurements of the molded epoxy discs are summarized in Table II. Halide concentrations were determined by aqueous extraction and are presented in Table III.

Semiconductor Grade Silicone-Epoxy Results

TGA. Weight loss measurements for the semiconductor grade silicone-epoxy formulations with and without FR are presented in Figure 3. For temperatures up to at least 325°C, no differences attributable to the FR are observed. Furthermore, below 275°, there is no difference for samples heated in air or nitrogen. Data were also obtained for a pure silicone molding compound and are presented in Figure 3 for comparison. Isothermal weight loss measurements revealed no difference between FR and non-FR silicone-epoxy compounds. After 12-days at 220°, a 20% weight loss was observed for the unfilled polymer, similar to the electrical grade novolac epoxy isothermal results.

TABLE II

X-RAY FLUORESCENCE MEASUREMENTS
Relative Total Br Concentration

Type	Sample Designation	Relative Br Concentration*
Electrical	A(FR)	1.0
Grade	B(1/2 FR)	0.6
Epoxy	C(no-FR)	0.0
Silicone-	D(FR)	1.1
Epoxy	E(no-FR)	0.0
Semiconductor	F(FR)	1.6
Epoxies	G(no-FR)	0.0

*uncorrected for attenuation in the compounds.

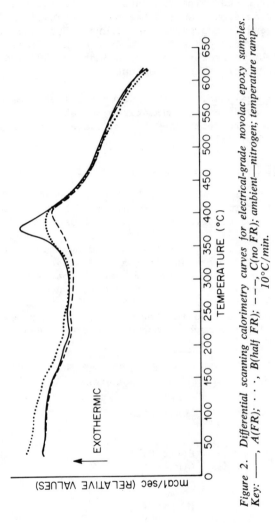

Figure 2. Differential scanning calorimetry curves for electrical-grade novolac epoxy samples. Key: ——, *A(FR);* · · · , *B(half FR);* – – –, *C(no FR); ambient—nitrogen; temperature ramp— 10°C/min.*

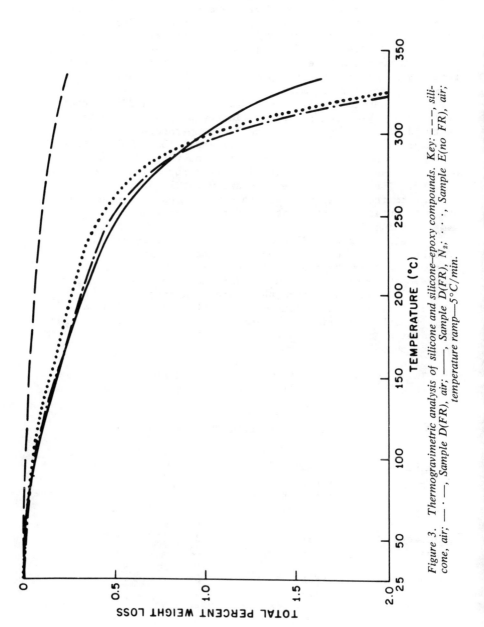

Figure 3.   Thermogravimetric analysis of silicone and silicone–epoxy compounds. Key: – –, sili-cone, air; – · –, Sample D(FR), air; ——, Sample D(FR), N_2; · ·, Sample E(no FR), air; temperature ramp—5°C/min.

EGA.   An isometric representation of the ion profiles
characteristic of the gases evolved from the silicone-epoxy
compounds are presented in Figures 4 and 5 for samples D(FR)
and E(no-FR), respectively. From these data the initial
weight loss observed in the TGA measurements of these samples
(50-150°C) is identified as being due to evolution of benzene
(m/e=78 parent ion; m/e39, 50, 51 and 52 fragment ions) and,
to a lesser extent, water vapor (m/e=17, 18). Contributions
from other species are minor.

The isometric ion plots of Figures 4 and 5 indicate that
evolution of benzene from the silicone-epoxy samples occurs
in two distinct stages, with the low temperature peak
attributable to residual solvent species. Above 200°C, thermal
degradation processes involving scission of the Si-phenyl
bond occur and account for the increased formation rate of
benzene. The other high temperature volatile products are
similar to those observed for the novolac epoxy samples[1], and
are attributed to decomposition of the epoxy fraction of
samples D and E.

Comparison of the silicone-epoxy ion profiles indicates
that the presence of the flame retardant in sample E has
little effect on the composition or formation rates of the
major volatile species. The specific ion profiles characteristic of HBr and CH_3Br from the flame retardant in sample E
are similar to those exhibited by the flame-retarded novolac
epoxy (sample A in Figure 1), and confirm the observation that
breakdown of the flame retardant has little effect on outgassing below 300°C.

DSC.   The principal feature of the DSC plots in Figure
6 for the FR and non-FR silicone-epoxies is the occurrence of

TABLE III

EXTRACTABLE HALIDES

Type	Sample Designation	Cl^-	Br^-
Electrical	A(FR)	650ppm*	160ppm*
Grade	B(1/2 FR)	650	120
Epoxy	C(no-FR)	650	---
Silicone-	D(FR)	<<25	<10
Epoxy	E(no-FR)	<<25	---
Semiconductor	F(FR)	65	10
Epoxy	G(no-FR)	200	---

---

*Aqueous extraction for 48-hrs. at 120°C
(ppm in molding compound).

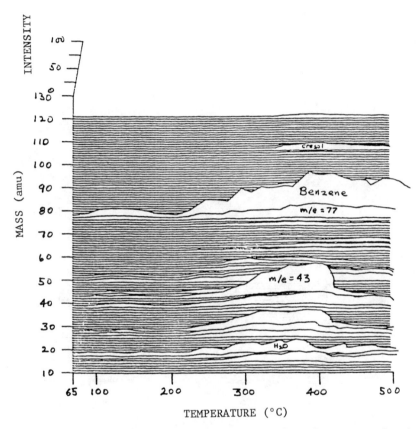

Figure 4. Isometric plots of mass spectral ion profiles silicone–epoxy Sample D(FR).

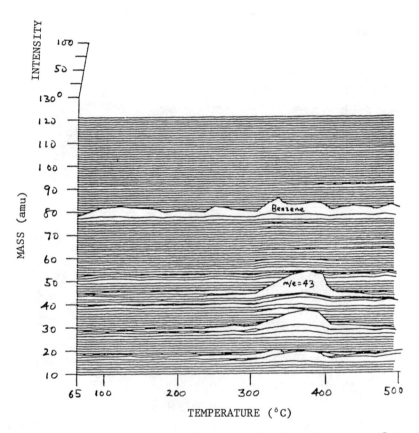

Figure 5.   Isometric plots of mass spectral ion profiles from silicone–epoxy Sample E(no FR).

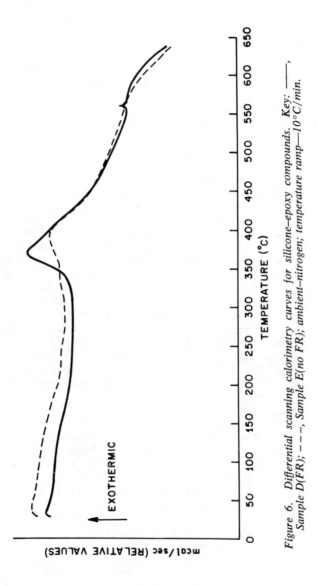

Figure 6. Differential scanning calorimetry curves for silicone–epoxy compounds. Key: ——, Sample D(FR); – – –, Sample E(no FR); ambient–nitrogen; temperature ramp—10°C/min.

an exotherm just below 400°C. As was observed for the
electrical grade epoxy samples, the exotherm is greater for
the FR compound. A nitrogen ambient was used to eliminate
possible interferences in the decomposition exotherms by
exothermic oxidation reactions. In runs at higher sensitivity,
no differences were observed in the DSC data for temperatures
below 300°C.

Elemental Analyses. X-ray fluorescence measurements
(Table II) indicate no major difference between the total Br
content in the FR silicone-epoxy (sample A). However,
extractable halide concentrations, listed in Table III, are
much lower for the silicone-epoxy samples.

Semiconductor Grade Epoxy Results

TGA. Unlike the previous two sets of molding compounds,
the semiconductor grade novolacs investigated in this section
do not differ solely in the presence or absence of a flame
retardant. However, they are considered to be relatively
equivalent. Weight loss measurements for the FR formulation,
sample F, and the non-FR compound, sample G, are presented
in Figure 7 for both nitrogen and air. Several differences
are observed in the weight loss curves for these samples.
First, below 300°C the weight loss in air for both compounds
is less than the loss in nitrogen. In fact, for the samples
heated in air a slight weight increase is recorded near 250°C.
This latter behavior indicates that oxidation of the epoxy
samples occurs with a rate that, at least initially, is faster
than weight loss through degradation and volatilization. However, this behavior is only transient since weight losses
greater than 20% are observed after isothermally heating these
samples in air at 220°C for 12-days.
Furthermore, the non-FR sample exhibits a lower thermal
stability in both air and nitrogen than the FR compound. This
was confirmed by the isothermal weight loss data, presented in
Figure 8, which indicate a significantly larger weight loss
for the non-FR epoxy compound. After 12-days this sample has
lost 70% more weight than the FR-epoxy (sample F). The
isothermal data for the FR-epoxy is very similar to that
characteristic of the electrical grade novolac epoxy and the
silicone-epoxy.
EGA. The overall ion profiles for the semiconductor grade
epoxy compounds are presented in Figures 9 and 10 for samples
F(FR) and G(no-FR), respectively. Very little outgassing is
observed from these samples below 200°C, in marked contrast to
the results obtained for the electrical grade epoxy samples[1].
These data clearly reflect the effects of the more stringent
processing controls employed in the production of the semiconductor grade materials. Also, because of the lower out-

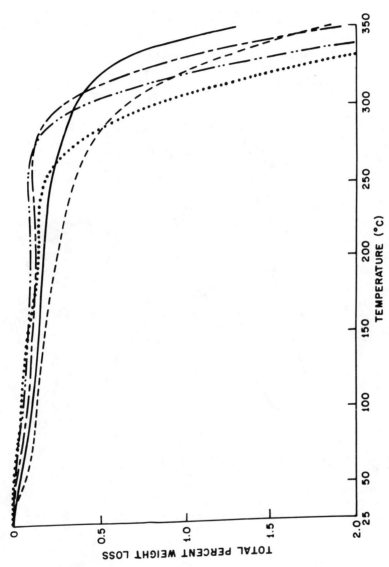

Figure 7.   Thermogravimetric analysis of semiconductor-grade novolac epoxy compounds. Key: (Sample G, no FR) – – –, N_2; · · ·, air; (Sample F, FR) ———, N_2; — · —, air (Run 1); — · · —, air (Run 2); temperature ramp—5°C/min.

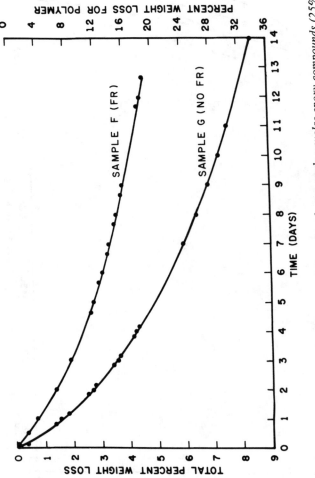

Figure 8.   Isothermal weight-loss curves for semiconductor-grade novolac epoxy compounds (25% polymer, 75% inert filler). Key: ambient—air, 220°C.

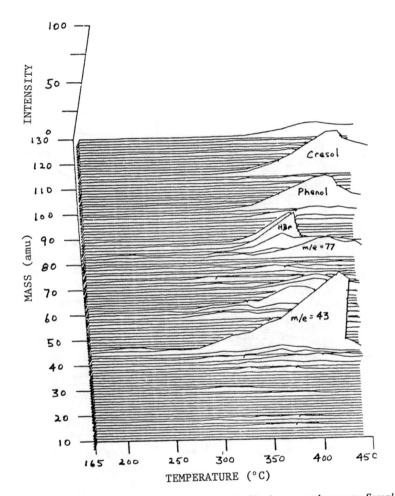

Figure 9.   Isometric plots of mass spectral ion profiles from novolac epoxy Sample F(FR).

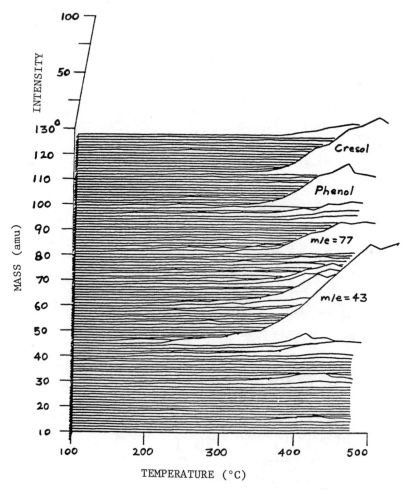

Figure 10.   Isometric plots of mass spectral ion profiles from novolac epoxy Sample G(no FR).

gassing, less interferences are present in the ion signals
characteristic of the volatile flame retardant species. Formation of HBr from the FR-epoxy (sample F) starting at 350°C is
clearly evident in the ion profile data of Figure 9. No
detectable products attributable to the flame retardant
are observed below 300°C.

DSC. Exotherms characteristic of the semiconductor grade
epoxy samples are presented in Figure 11. The low temperature
exotherm at 330°C is unique to the FR compound. However, since
these two epoxy compounds are not strictly equivalent, as
mentioned earlier, the appearance of this exotherm cannot
be unequivocally attributed to the presence of the flame
retardant. The high temperature exotherm above 350°C, on the
other hand, is observed in both samples, with the FR-compound
exhibiting a larger peak. This behavior is identical to that
observed for the electrical grade novolac epoxy (Figure 2)
and the silocone-epoxy (Figure 6), and, in this case, is
attributed to reactions involving the flame retardant. Again,
no significant differences are observed in the DSC plots
below 300°C.

Elemental Analyses. The X-ray fluorescence measurements
of Table II indicate a somewhat higher total Br content for the
semiconductor grade novolac (sample F) than for the electrical
grade (sample A). However, the data of Table III indicate
halide concentrations in the extract that are an order of
magnitude lower for the semiconductor grade novolac (sample G)
is a factor of three material (sample F).

Discussion

Electrical/Electronic Grade Epoxy. TGA and DSC analyses
revealed no difference in thermal degradation below 200°C due
to the presence of FR. DSC and EGA measurements showed that
the FR breaks down above 350°C, in the range where it can perform its designated function. However, the EGA analysis did
detect a small quantity of bromine-containing fractions below
200°C, and aqueous extraction revealed a fairly high Br⁻
concentration of 160 ppm.
Since no difference was found between FR and non-FR
formulations in device aging studies[2], some other cause must
account for the relatively early failures observed[2,3] for
devices molded in the electrical grade epoxy material and
aged under bias at 200°C. These failures are attributed to
chloride contamination present in the non-semiconductor grade
epoxy resin. The extractable Cl⁻ concentration is a factor
of four higher than Br⁻, and this is correlated with a much
higher concentration of CH_3Cl than CH_3Br in the EGA data below
200°C[1]. The high Br⁻ concentration is also attributed to the

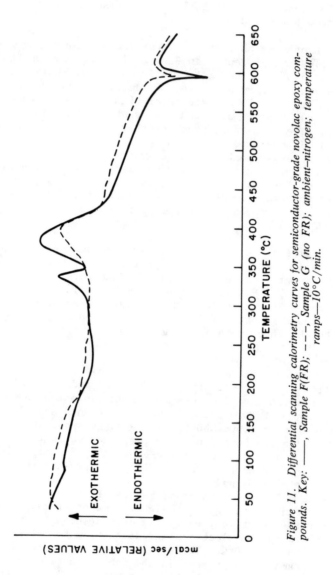

Figure 11. Differential scanning calorimetry curves for semiconductor-grade novolac epoxy compounds. Key: ———, Sample F(FR); – – –, Sample G (no FR); ambient—nitrogen; temperature ramps—10°C/min.

to the poorer "housekeeping" in preparation of a non-semiconductor grade compound, rather than to an inherent property of a FR epoxy. It should be noted that even this non-semiconductor grade novolac was found to have adequate reliability for molded devices operated below 100°C[2].

Semiconductor Grade Silicone-Epoxy. TGA, DSC, and EGA analyses revealed no difference between the FR and non-FR compounds below 200°C. The FR moieties again decomposed only in the temperature range above 350°C. There was very little Cl^- or Br^- in the aqueous extract, and no CH_3Cl or CH_3Br was detected in the EGA product profiles. This shows the capability of material formulators to supply very clean semiconductor grade molding compounds.

Nonetheless, the FR silicone-epoxy has yielded significantly lower reliability than the equivalent compound without FR in bias aging of devices at 200°C[3]. In those experiments, both the activation energy and the electrical failure mode were indicative of failures due to cation migration. This is in marked contrast to the activation energy and failure anlaysis results reported for the electrical grade epoxy[2], which showed device failures due to corrosion of metallization, presumably due to Cl^-. Therefore, device failure for the FR silicone-epoxy is, again, not directly related to the presence of Br. However, it appears that the flame retardant system may be responsible for cation generation via some mechanism as yet unexplained.

Semiconductor Grade Epoxies. As was the case for the semiconductor grade silicone-epoxy, there was no difference between FR and non-FR epoxies recorded by either DSC or EGA below 200°C. However, the nominally equivalent non-FR epoxy exhibited significantly lower thermal stability as indicated by the isothermal TGA data. Furthermore, the aqueous extract of the non-FR compound contained more than twice as much Cl^- as the combined concentrations of Cl^- and Br^- in the FR epoxy. Although there have been no direct comparisons on device aging with these two epoxies, the above findings indicate that the FR compound, being cleaner and more thermally stable, could actually be the better material for encapsulation applications.

Acknowledgments

The authors thank A. Zabotti for thermal analysis data, R. J. Holmes for X-ray fluorescence measurements, and T. W. Zuber for molding.

Literature Cited

1. Lum, R. M.; Feinstein, L. G. Microelectronics and Reliability, 1981, 21, 15.
2. Feinstein, L. G. Microelectronics and Reliability
   1981, 21, 00.
3. Masessa, A. J.; Feinstein, L. G. To be published.
4. Sherman, S.; Gannon, T.; Buchi, G.; Howell, W. R. in
   "Kirk-Othmer: Encyclopedia of Chemical Technology",
   Vol. 9, 3rd. edit., John Wiley and Sons, 1980, pp. 267-290.
5. Helfand, D.; Villani, T. Proc. 14th. Elec./Electr. Conf.,
   1979, pp. 290-297.
6. Melliar-Smith, C. M.; Matsuoka S; Hubbauer, P. private
   communication.
7. Lum, R. M. J. Polym. Sci. Chem. Ed. 1979, 17, 203.
8. Davidson, T. E.; Roberts, C. W. J. Appl. Polym. Sci.
   1980, 25, 1491.

RECEIVED November 3, 1981.

Electrical Switching and Memory Phenomena in Semiconducting Organic Thin Films

R. S. POTEMBER and T. O. POEHLER

Johns Hopkins University, Applied Physics Laboratory, Laurel, MD 20810

Switching and Memory Device: Materials and Fabrication

This paper is a report on stable and reproducible current-controlled bistable electrical switching and memory phenomena observed in polycrystalline metal-organic semiconducting films. The effects are observed in films of either copper or silver complexed with the electron acceptors tetracyanoethylene (TCNE), tetracyanonapthoquinodimethane (TNAP), tetracyanoquinodimethane (TCNQ), (1) or other TCNQ derivatives shown below. The character of the switching in going from a high- to a low-impedance state in these organic charge-transfer complexes is believed to be comparable in many respects to existing inorganic materials. The basic configuration of the device, shown in Figure 1, consists of a 5-10 μm thick polycrystalline aggregate of a copper or a silver charge-transfer complex sandwiched between two metal electrodes. Electrical connection is made to the two metal electrodes through silver conducting paste or through liquid metals of mercury gallium or gallium-indium eutectic. Fabrication of the device consists of first mechanically removing any oxide layers and organic contaminants from either a piece of copper or silver metal foil. The cleaned metal foil is then placed in a solution of dry and degassed acetonitrile which has been saturated with a neutral acceptor molecule, for example, TCNQ°. The neutral acceptors used in all of these experiments are recrystallized twice from acetonitril and then sublimed under a high vacuum prior to their use. (2) When the solution saturated with the neutral acceptor is brought in contact with a metal substrate of either copper or silver, a rapid oxidation-reduction reaction occurs in which the corresponding metal salt of the ion-radical acceptor molecule is formed. The basic reaction is shown in Equation 1 for copper and TCNQ°.

This technique of forming semiconducting films by direct oxidation-reduction is used to grow highly microcrystalline films directly on the copper or silver substrate. These films show a metallic sheen and can be grown to a thickness of 10 μm in a

0097-6156/82/0184-0233$05.00/0

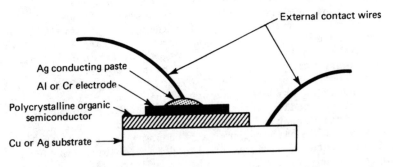

TCNE

TCNQ

TNAP

$R_1 = -H$, $R_2 = -H$
$R_1 = -CH_3$, $R_2 = -H$
$R_1 = -OCH_3$, $R_2 = -H$
$R_1 = -F$, $R_2 = -F$

External contact wires

Ag conducting paste
Al or Cr electrode
Polycrystalline organic semiconductor
Cu or Ag substrate

Figure 1.   Schematic of an organic switching device.

$Cu^0 +$ TCNQ $\rightleftharpoons Cu^+$

Equation 1.

matter of minutes. Once the polycrystalline film has been grown
to the desired thickness, the growth process can be terminated by
simply removing the metal substrate containing the organic layer
from the acetonitrile solution; this terminates the redox reaction. The two component structure is gently washed with additional acetonitrile to remove any excess neutral acceptor molecules and dried under a vacuum to remove any traces of solvent.
Elemental analysis performed on polycrystalline films of Cu-TCNQ
and Cu-TNAP removed from the copper substrate reveals that the
metal/acceptor ratio is 1:1 in both complexes. (3) Finally, the
three component structure is complete when a top metal electrode
of either aluminum or chromium is evaporated or sputtered directly
on the organic film.

Electrical Behavior

Threshold and memory behavior is observed in these materials
by examining current as a function of voltage across the two
terminal structure. Figure 2 shows a typical dc current-voltage
curve for a 3.75 μm thick Cu/Cu-TNAP/Al system. The trace in
Figure 2, as well as all other I-V measurements presented in this
paper, are made with a 10^2-Ω load resistor in series with the
device. Figure 2 shows that there are two stable non-ohmic
resistive states in the material. These two states, labeled "OFF"
state and "ON" state, are essentially insensitive to moisture,
light, and the polarity of the applied voltage. A rapid switching
is observed from the "OFF" to the "ON" state along the load line
when an applied field across the sample surpasses a threshold
value (V_{th}) of 2.7 V. This corresponds to a field strength of
approximately 8.1×10^3 V/cm. At this field strength the initial
high impedance of the device, 1.25×10^4 ohms, drops to a low impedance value of 190 ohms. This rise in current to 4 ma and con-
current decrease in the voltage to approximately 1.2 V along the
load line is observed in the Cu-TNAP system. It is representative
of the switching effects observed in all of the metal charge-
transfer salts examined and is characteristic of all two terminal
S-shaped or current-controlled negative-resistance switches. (4)
In addition, it has been observed in all of the materials
investigated that once the film is in the "ON" state it will
remain in that state as long as an external field is applied. In
every case studied, the film eventually returned to its initial
high-impedance state after the applied field was removed. It was
also found that the time required to switch back to the initial
state appeared to be directly proportional to the film thickness,
duration of the applied field, and the amount of power dissipated
in the sample while in this state.
Three general trends are noted in the "ON" state character
of the copper and silver complexes as related to the different
acceptor molecules. The first is that the copper salts consis-
tently exhibited greater stability and reproducibility over the
corresponding silver salts of the same acceptor. Second, it is

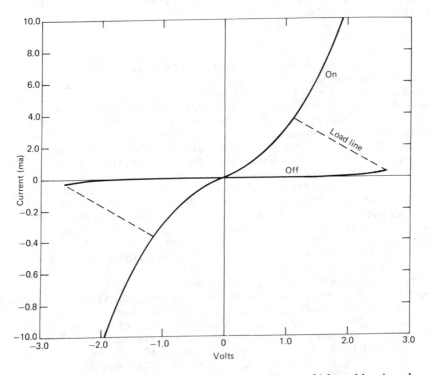

Figure 2.   Typical dc current–voltage characteristic showing high- and low-imped-
ance states for a 3.75-μm Cu–TNAP sample.

possible to correlate the preferred switching behavior of the different complexes to the reduction potential of the various acceptors. This plot is shown in Figure 3 using copper as a donor in each case. It appears that for devices made from weak electron acceptors, the switching behavior is usually of the threshold type, i.e., when the applied voltage is removed from a device in the "ON" state, the device will immediately return to the "OFF" state. On the other hand, for strong electron acceptors a memory effect is observed. This memory state remains intact from a few minutes up to several days and can often be removed by the application of a short pulse of current in either direction. For intermediate strength acceptors, it is possible to operate the device as either a memory switch or a threshold switch by varying the strength or the duration of the applied field in the low-impedance state. Third, it also recognized that the field strength of the switching threshold tends to parallel the strength of the acceptor. For instance, the copper salt of $TCNQ(OMe)_2$ switches at a field strength of approximately 2×10^3 V/cm, while the copper salt of $TCNQF_4$ is found to switch at a field strength of about 2×10^4 V/cm. It is clear that these three trends are related to the reduction potential of the acceptor calculated from solution redox potentials (5). However, as these values do not always parallel the values found in the solid phase, a more quantitative description relating to the switching behavior to the acceptor cannot be made unless the various contributions to the binding energy of the different ion-radical salts are considered.

The response to a very short pulse is exemplified in the next figure. Figure 4 is an oscilloscope trace showing both the leading edge of a voltage pulse and current pulse versus time for a Cu-TNAP sample in response to a rectangular voltage pulse with a 4 nsec rise time. This voltage pulse switched the sample from the high- to the low-impedance state and contained a 1.0 V overvoltage to eliminate any current oscillations between the "OFF" and "ON" states. Current oscillations arise when the applied voltage is set very close to V_{th}. It is not possible from this experiment to determine values for the conventional delay times and rise times because the combined delay and rise times appear to be less than 4 nsec (the limiting rise time of the pulse generator). This experiment suggests that the mechanism of the switching phenomena is not due to thermal effects (6) which have been used to describe switching and memory phenomena in many other systems. From Figure 4, it appears that the delay time is shorter than reported values for inorganic semiconductors under the same experimental conditions. A recent example of delay times in an inorganic amorphous material is given for the composition $Te_{10}As_{35}Ge_7Si_{17}P_1$ (7), approximately 1 μm thick, sandwiched between two molybdenum electrodes. A typical delay time reported for this device in response to a single 12 V pulse is about 2 μsec. To reduce the delay time to a value of 10 nsec, a 30 V pulse (18 V overvoltage) was required.

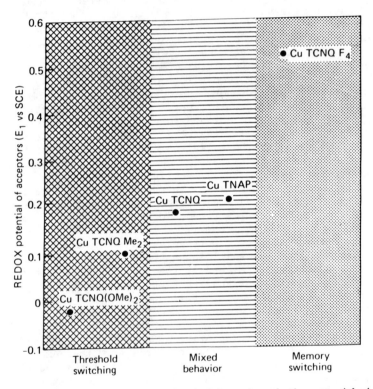

Figure 3. Type of switching behavior plotted vs. the reduction potential of the acceptor.

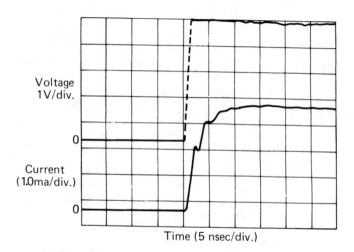

Figure 4.   Transient response to a 4-ns rise time rectangular pulse.

An experiment was designed to determine if the device generates an open-circuit voltage or electromotive force (emf) when
returning from the low- to the high-impedance mode. The appearance of a spontaneous emf (8) would indicate that an electrochemical reaction was responsible for switching phenomena. In this
experiment: 1) an applied voltage in excess of the threshold
voltage was used to place a Cu-TNAP sample into a low-impedance
state where it would remain for a short time after the applied
voltage was removed, i.e., memory state; 2) the sample was then
externally short-circuited to eliminate any capacitive effects,
and finally; 3) a high input impedance storage oscilloscope was
used to measure open-circuit discharge voltage when the sample
spontaneously returned to its original high-impedance state. The
oscilloscope was set to trigger whenever a voltage exceeding a
few millivolts appeared across the sample. The results are shown
in Figure 5 where the spontaneous open-circuit voltage measured
by the oscilloscope is reproduced and is seen to have a maximum
voltage discharge of approximately 0.3 volts.

The electromotive force (emf) of 0.3 volts observed in this
experiment does show that the mechanism by which the switching
occurs is consistent with a field induced solid-state reversible
electrochemical reaction associated with the metal charge-transfer
salts.

Infrared Reflectance Spectra of Cu-TCNQ Semiconducting Films

To investigate the formal charge of TCNQ in the semiconducting films of Cu-TCNQ, the infrared reflectance spectra was
recorded at room temperature for crystalline Cu-TCNQ films before
and after an external electric field was applied to the sample.
The applied field in this experiment was of a strength comparable
to that in switching device structures, i.e., a field in excess
of 10^4 V/cm was used. The results were then compared to the
reflection spectra measured for other crystalline metal-TCNQ
radical-anion salts. These salts are known to exist as either
simple or complex salts in the solid-state. The crystalline
materials investigated were lithium-TCNQ, cesium-TCNQ, copper-
TCNQ (prepared by a metathetical reaction) and copper-TCNQ grown
on copper substrates in the manner similar to the switching
devices. Specifically, the region of the infrared spectrum
measured was between 2000 to 2500 cm^{-1} (0.25 to 0.3 eV). This
spectral region corresponds to the ν_2 C≡N stretching mode in
TCNQ. Previous studies have provided evidence to link the frequency assignment of C≡N stretching and C=C stretching modes to
the degree of charge transfer in complexes of TCNQ. (9) In these
investigations a frequency shift to lower energy is reported as
charge density increases on TCNQ.

The Cu-TCNQ switching material was subjected to electric
fields by clamping a thin highly insulating film of either teflon
or polyethylene between the surface of the Cu-TCNQ film on a cop-

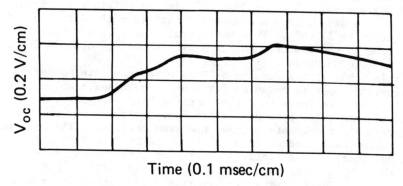

Figure 5.   *Spontaneous open-circuit potential generated in a Cu/Cu–TNAP/Al sample at room temperature.*

per substrate and an external top metal electrode. The reflectance spectrum was recorded after removing the field and separating the Cu-TCNQ (on the copper substrate) from the top electrode and the insulating plastic film. All of the samples were freshly prepared and the solid-state diffuse reflectance spectra was recorded on a Perkin-Elmer 621 Grating IR Spectrometer. Wherever possible, elemental analysis was performed on the samples to verify their composition.

The upper trace in Figure 6 is a reflectance spectrum of a crystalline film of Cu-TCNQ before the application of an electric field. A moderately strong infrared active mode for CN is observed to dominate the region characterized by a single line center at approximately 2320 cm^{-1}. The lower trace (Figure 6) is a reflectance spectrum of the same film of Cu-TCNQ, but in this spectrum an electric field has been applied to the sample for 72 hours. In this trace there are two reflectance maxima. One line can be assigned a value of 2321 cm^{-1} which is nearly identical to the maximum value seen at 2320 cm^{-1} in the original spectrum. However, a second line has appeared that is shifted to a higher frequency by 21 cm^{-1}. This additional peak is indicative of a decrease in the electron charge on the CN moiety of some fraction of the TCNQ molecules. (10)

In Table I, the results of this experiment are compared to reflectance spectra measured for other simple and complex metal-TCNQ salts. We found that the CN stretching mode in reflectance measurements shifted to higher frequency by about 100 cm^{-1} from absorption measurements made on the same material. The peak in the reflectance band at 2320 cm^{-1} for the Cu-TCNQ film prior to the application of a field is consistent with the values measured for the simple (1:1) salts of Li$^+$(TCNQ$^\mp$) and Cu$^+$(TCNQ$^\mp$) tabulated in Table I. These crystalline materials are simple salts which do not contain neutral TCNQ°. On the other hand the spectra of a Cu-TCNQ film after the application of an applied field closely resembles the spectra of Cs$_2$(TCNQ$^\mp$)$_3$ with two CN stretching modes separated by ~ 20 cm^{-1}. Cs$_2$(TCNQ$^\mp$)$_3$ is a complex salt which contains neutral TCNQ° and radical-anion TCNQ$^\mp$. (11)

The diffuse reflectance spectra reported in Table I show that it is possible to assign a CN stretching frequency to both neutral and radical-anion TCNQ in crystalline samples of metal-TCNQ complexes because the reflectance peak for neutral TCNQ is shifted ~ 20 cm^{-1} higher in frequency than for radical-anion TCNQ$^\mp$. Specifically, the reflectance data for Cu-TCNQ when compared to other metal-TCNQ salts of known composition strongly suggests that neutral TCNQ° is not present in the unswitched Cu-TCNQ films. On the other hand, the additional peak that appears in the spectra of Cu-TCNQ subjected to an applied field shows a peak superimposable with the peak recorded for neutral TCNQ° in Cs$_2$(TCNQ$^\mp$)$_3$. This evidence suggests that neutral TCNQ° is formed in a solid-state field induced phase transition when electric fields are applied to crystalline films of Cu-TCNQ grown on copper substrates.

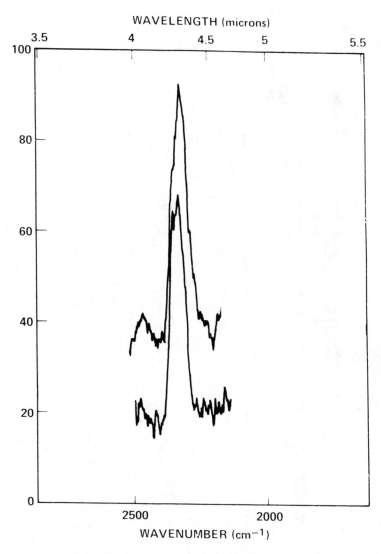

Figure 6.   Reflectance spectra of a crystalline film of Cu–TCNQ on copper before and after the application of an electric field.

Table I. Comparison of Reflectance Maximum for the CN Stretching Mode in TCNQ for Various Metal–TCNQ Salts.

TCNQ SALT	COMMENTS	REFLECTION MAXIMUM (cm^{-1})
Li TCNQ	SIMPLE 1:1 SALT	2320
Cs_2 $TCNQ_3$	COMPLEX 2:3 SALT	2322 AND 2344
Cu TCNQ	SIMPLE 1:1 SALT PREPARED BY METATHETICAL REACTION	2323
Cu TCNQ SWITCH	BEFORE APPLICATION OF ELECTRIC FIELD	2320
Cu TCNQ SWITCH	AFTER APPLICATION OF ELECTRIC FIELD	2321 AND 2340

Conclusions

It is postulated that mixed-valence species or complex salts (12) formed as a result of this field induced redox reaction control the semiconducting behavior of these films and these complex salts exist in a solid-state equilibrium with the simple 1:1 salt. Since non-integral oxidation states are common in solids, it is difficult to predict exact stoichiometry in the equilibrium equation, but a likely equation for switching in Cu-TCNQ, for example, may involve

$$[Cu^+(TCNQ^{\overline{\cdot}})]_n \rightleftarrows Cu^o_x + [Cu^+(TCNQ^{\overline{\cdot}})]_{n-x} + (TCNQ^o)_x.$$

In addition, an ionic or a molecular displacement associated with this equilibrium would explain the observed memory phenomena and the fact that all the devices show only two stable resistive states.

Since conduction in these narrow band semiconducting salts of TCNQ is believed to be limited by the motion of unpaired electrons along the stacks of TCNQ molecules, this interpretation is in accordance with the electrical behavior reported in these films when fabricated into switching devices. (13, 14, 15) In a simple salt like $Cu^+(TCNQ^{\overline{\cdot}})$ there is roughly one unpaired electron per molecule which tends to keep electrostatic repulsion in the ground state configuration at a minimum. The low conductivity reported in these simple salts is due in part to an increase in the energy required to overcome the repulsive coulomb forces that result when a conduction electron is removed from one $TCNQ^{\overline{\cdot}}$ and placed into a higher energy orbital of another $TCNQ^{\overline{\cdot}}$ molecule.

In the case of a mixed-valence salt containing neutral $TCNQ^o$ there are more TCNQ molecules than there are unpaired electrons and, therefore, electrostatic repulsion of charge carriers is kept at a minimum by allowing conduction electrons to occupy the empty molecular orbitals of $TCNQ^o$. This is a lower energy pathway compared to putting more than one electron on the same TCNQ site and it may explain how mixed-valence semiconducting salts like Cs_2-$(TCNQ^{\overline{\cdot}})_3$ and the "switched" form of Cu-TCNQ can exhibit greater conductivity than similar salts with 1:1 stoichiometry.

Acknowledgments

We gratefully acknowledge support by the National Science Foundation (DMR 80-15318) and the Department of the Navy (N00024-81C-5301).

Literature Cited

1. Potember, R. S.; Poehler, T. O.; Cowan, D. .O. App. Phys. Lett. 1979, 34, 405.
2. Gemmer, R. V.; Cowan, D. O.; Poehler, T. O.; Bloch, A. N.; Pyle, R. E.; Banks, R. H. J. Org. Chem. 1975, 40, 3544.
3. Elemental analysis was performed by Galbraith Laboratories, Inc., Knoxville, Tennessee 37291.
4. Owen, A. E.; Robertson, J. M. IEEE Trans. Electron Devices 1973, 20, 105.
5. Values for the reduction potential of acceptor were taken from Wheland, R. C.; Gillson, J. L. J. Am. Chem. Soc. 1976, 98, 3916.
6. Buckley, W. D.; Holmberg, S. H. Solid-State Electron. 1975, 18, 127.
7. Reinhard, D. K. App. Phys. Lett. 1977, 31, 527.
8. A spontaneous electrochemical reaction is reported in magnesium-TCNQ salts. See Gutmann, F.; Herman, A. M.; Rembaum, A. J. Electrochem. Soc. 1967, 114, 323.
9. Matsuzaki, S.; Kutwata, R.; Toyoda, K. Solid State Commun. 1980, 33, 403.
10. Khatkale, M. S.; Devlin, J. P. J. Chem. Phys. 1979, 70, 1851.
11. Fritchie, C. J.; Arthur, Jr., P. Acta Cryst. 1966, 21, 139.
12. For a discussion of complex TCNQ salts see LeBlanc, Jr., O. H. J. Chem. Phys. 1965, 42 4307.
13. Torrance, J. B.; Scott, B. A.; Kaufman, F. B. Solid State Commun. 1975, 17, 1369.
14. Soos, Z. G. Ann. Rev. Chem. 1974, 25, 121.
15. Hubbard, J. Phys. Rev. 1978, B17, 494.

RECEIVED November 16, 1981.

INDEX

INDEX

Jacket design by Kathleen Schaner.
Production by Susan Moses and V. J. DeVeaux.

Elements typeset by Service Composition Co., Baltimore, MD.
Printed and bound by The Maple Press Co., York, PA.